COULTER LIBRARY ONONDAGA COMM. COLL.
TH9145.W45
Whitman, Lawrence E. Fire prevention

3 0418 00030220 6

D1793475

TH
9145　　Whitman, Lawrence E.
W45
　　　　　FIRE PREVENTION

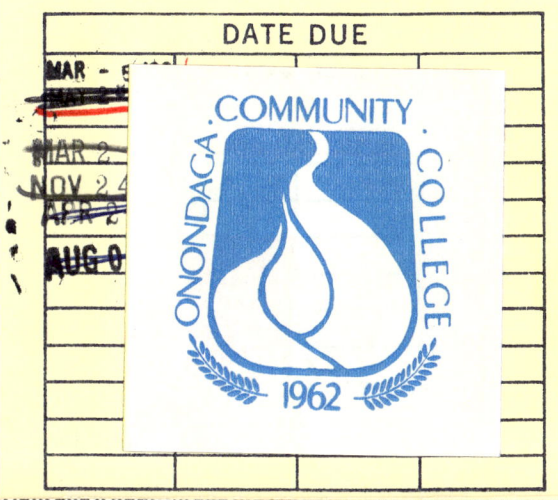

DATE DUE

MAR -
MAR 2
NOV 24
APR 2
AUG 0

0165　01 154029　01 6　(IC=0)
WHITMAN, LAWRENCE E.
FIRE PREVENTION /
(0) [1979

TH 9145 .W45

THE SIDNEY B. COULTER LIBRARY
ONONDAGA COMMUNITY COLLEGE
RTE 173, ONONDAGA HILL
SYRACUSE, NEW YORK 13215

Fire

Prevention

Lawrence E. Whitman

Nelson-Hall nh Chicago

Library of Congress Cataloging in Publication Data

Whitman, Lawrence E.
    Fire prevention.

    Bibliography: p.
    Includes index.
    1. Fire prevention. I. Title.
TH9145.W45   628.9'22   78-26894
ISBN 0-88229-359-1

Copyright © 1979 by Lawrence E. Whitman

All rights reserved. No part of this book may be reproduced in any form without permission in writing from the publisher, except by a reviewer who wishes to quote brief passages in connection with a review written for broadcast or for inclusion in a magazine or newspaper. For information address Nelson-Hall Inc., Publishers, 111 North Canal Street, Chicago, Illinois 60606.

Manufactured in the United States of America

10   9   8   7   6   5   4   3   2   1

Contents

Preface
The Fire Prevention Problem ix
   chapter 1
History and Philosophy of Fire Prevention 1
        Fire Prevention Begins in the Ancient World/
        Fire Prevention in Medieval Europe/Early American and
        Fire Safety/Advances in the Nineteenth Century/Fire
        Prevention and the Technological Revolution/Some Errors
        in Current Fire Prevention Thinking/A Look into the
        Future/Fire Prevention Philosophy
   chapter 2
Fire Prevention Organizations and Their Functions 11
        Federal/State/Municipal/Private
   chapter 3
Building Construction 21
        Building Codes and Ordinances and Their Uses/Code
        "Blind" Spots/Major Types of Construction, Old and New/
        Hazards for Buildings Under Construction or Undergoing
        Major Alterations/Horizontal and Vertical Fire Cutoffs/
        Concealed Spaces/Interior Finish/Interpreting Fire Tests of
        Building Construction and Materials/Effects of Major
        Occupancy Changes on Building Fire Safety/Special Life
        Safety Requirements for Nursing Homes, Hospitals, Schools,
        and Similar Institutions/High-Rise Buildings/Unusual
        Types of Buildings

vi   CONTENTS

chapter 4
Fire Exposure 49
        External Exposures/Internal Exposures
chapter 5
Building Utility Systems 65
        Definitions/Electrical Systems/Air-Handling Systems/
Gas Systems/Heating Systems
chapter 6
Fire Protection Systems 97
        Automatic Sprinkler Systems/Special Protection Systems
chapter 7
Occupancy Utility Systems 125
        Electrical Systems/Air-Handling Systems
chapter 8
Industrial Furnaces, Ovens, Driers, and Incinerators 141
        Process Furnaces/Ovens and Driers/Incinerators
chapter 9
Flammable Liquids and Gases 153
        Flammable Liquids/Flammable Gases
chapter 10
Special Chemical Hazards 185
        Air- and Water-Reactive Chemicals/Oxidizing Agents/
Corrosive Chemicals/Toxic Chemicals/Explosives/Cryogenics
Unstable Chemicals
chapter 11
Arson Detection and Investigation 221
        The Problem/Motivations for Arson/Arson Law/Arson
Investigations/Preparing Evidence
chapter 12
Plastics 239
        Uses/Nitrocellulose/Basic Classifications/Creating
Plastics/Finished Product Hazards/Fire Tests/Some
Instructive Fires
chapter 13
General Occupancy Problems 251
        Housekeeping/Key Operations/Plant Shutdowns/
Disaster Recovery/Plant Management and Organization
chapter 14
Fire Loss Investigations 263
        General Investigation Procedures/The Report Form/
Some Investigative Techniques

chapter 15
Public Relations and Education 281
Public Relations/Education
chapter 16
Fire Prevention Inspections 291
Goals/Attitudes/Inspection Frequencies/Preparation for Inspections/Inspections/Exit Interviews/Reports
appendix A
Metric Conversion Table 307
appendix B
Other Useful Measurements 307
appendix C
Useful Formulas 307
Selected Reference Bibliography 308
Index 309

The Fire Prevention Problem

Fire is a serious national problem, the true nature of which is not understood by most Americans. At least in part, the lack of understanding is due to the absence of a clear-cut picture of the real extent of our annual fire losses and their effect on the people who do not experience a fire personally.

Each year, statistics are published on the number of fire deaths and the amount of direct property damage. Never, however, has the author seen realistic published information on the other costs of fire —for example, lost productive man-years, lost jobs, business failures, hospital costs, welfare costs, and insurance costs. Facts on these phases of fire losses are beginning to emerge, and the picture presented is not a pretty one. For example, unless fire prevention practices are upgraded, we can expect the following losses.

1. During the time it takes to read this book, fire will kill at least 3 Americans, injure 100 others, of whom 16 will be hospitalized for two or more weeks; and destroy half a million dollars worth of property. Of the injured, several will be crippled or disfigured for life.

2. Each year, various fire-related costs—for example, direct property losses, lost man-years of production, lost jobs, business failures, medical treatment, and welfare payments—will take approximately $50 billion out of American pocketbooks through increased taxes, increased insurance costs, and other levies. This loss is roughly equal to the total income of all the residents of Virginia for two years.

3. The annual direct property losses alone will be equivalent to the loss of all the assets of an organization the size of Bethlehem Steel Corporation.

4. Each year, the number of people burned seriously enough to require hospitalization will about equal the population of Richmond, Virginia.

If the American fire experience were similar to that of other industrial nations, or if there were any consistency in the fire pattern across the nation, it might be possible to write off the losses as part of the price of progress. Unfortunately, the nation as a whole burns people and property from 2.5 to 29 times as rapidly as other countries or our better-run public and private organizations.

When we consider that generally America surpasses other nations in quality of building construction, fire protection systems, and staffing, training, and equipping of fire departments, it becomes apparent that the American fire loss record is a disgrace. The record appears particularly shameful when one realizes that a few American organizations of national scope consistently compile records far superior to those of most of the industrialized nations.

There is only one possible explanation of the facts just cited. Nations, public agencies, and private companies with outstanding fire records understand what fire prevention really is and put their knowledge to work.

Most Americans are indifferent to the fire problem because they do not realize how a fire a thousand miles away—or, for that matter, in the next city—can affect them in the form of increased taxes, higher insurance rates, and loss of jobs. People do not really understand fire prevention and the vital part that individual efforts can play in saving lives, property, and jobs. We who are in the fire service have the responsibility for correcting such public indifference and misunderstanding. A nation that can put man on the moon can conquer fire. Set the goal, show the way, and the people will do the rest.

This book has two purposes (1) to introduce fire science students to the scope of fire prevention and to the role of prevention in an overall fire program, and (2) to provide a better understanding of the ways in which good and poor programs may affect both nearby and remote communities.

A book of this size can only serve as a basic introduction to a vitally important subject. It is hoped that, by providing illustrations and closing some gaps in general knowledge, the book will lead the reader

Preface

to seek out more detailed information on various types of hazards and means to correct deficiencies.

The opinions expressed in the book are the sole responsibility of the author. Correspondence from students or others using the text who wish to offer comments or suggestions for correction or improvement would be appreciated.

The National Fire Protection Association has recently changed its procedures for gathering statistics on fire deaths and injuries. These changes, however, affect only the numbers of reported casualties. They do not affect the conclusions to be drawn from the record.

chapter 1
History and Philosophy of Fire Prevention

No organization, whatever its purpose, and no method for managing an organization ever came into being without having roots in the past. Fire prevention organizations, large or small, are no exception to this general rule. Therefore, if we are to have a realistic understanding of where fire prevention stands today and the directions it should take tomorrow, we need to know something about its past history.

Fire Prevention Begins in the Ancient World

Recorded history only goes back about six thousand years. During that time, some twenty or twenty-one civilizations developed, and each had fire problems in one form or another. Reactions to these problems (in part fire prevention history) differed widely, depending upon the stage of civilization reached.

Perhaps the earliest reference to fire control is in the law code of Hammurabi, a Babylonian king of the twentieth century B.C. As far as the author is aware, the code does not specifically mention fire, but it does provide for the execution of persons responsible for the destruction of inhabited buildings. (Incidentally, this code provision was also a handy means for getting rid of potential enemies.)

During the next one thousand six hundred years, the civilized world devised few changes in its approach to fire problems except for the gradual introduction of unorganized attempts to control fire with whatever tools were available. That innovation was not much, but it was a beginning, and out of it came the first major step forward, the formation of organized fire brigades in Rome in about 300 B.C. These brigades were composed of slaves directed by untrained overseers. Predictably—from our point of view—their efforts were of limited value.

Recognizing the weaknesses of the disorganized and untrained slave brigades, Roman authorities near the beginning of the Christian Era established a paid department, which ultimately had several thousand men. They were provided with primitive fire control tools and received limited training in their use. These men patrolled streets and had authority to administer corporal punishment to offenders of the primitive fire codes then being developed. To many people, this development marks the real beginning of fire prevention. Unfortunately, before this organization could develop into an effective fire prevention system, Rome fell, and the Dark Ages descended on Europe. Fire prevention then all but disappeared for several hundred years.

Fire Prevention in Medieval Europe

As Europe began to emerge from the Dark Ages, the fire problem again began to register on the minds of responsible individuals. With little background except vague memories of the vanished Roman firefighting system, and with no real knowledge of fire behavior, they began to take faltering steps towards putting fire under control.

Perhaps not the first step, but certainly the most important early step, was taken in Oxford, England, in 872 A.D. when a curfew was established and patrols were organized to enforce it. As then used, the word *curfew* meant "cover or extinguish fire." In Oxford, all unattended fires were required to be extinguished when the curfew bell was rung in the evening. This practice spread, and, by the time of the Norman conquest in 1066 A.D., was general throughout England. Since most English homes of the day had wood chimneys daubed with mud on the inside, and thatched roofs, this simple fire prevention measure undoubtedly saved many buildings and many lives.

During the years following the establishment of curfew regulations, the English gradually gained some knowledge of fire behavior. As a result, the city of London during the twelfth century enacted ordinances requiring new buildings to have stone walls and slate or tile roofing. There were, however, few provisions for enforcement and none covering the thousands of existing combustible buildings. As far as the author has been able to determine, the ordinances had no realistic requirements concerning building spacing, piercing of the stone walls for windows and doors, size of buildings, or permissible occupancies. These ordinances, therefore, were of limited effectiveness, but they were a step forward.

Some four hundred years after the London ordinances were enacted, the British Parliament passed an act prohibiting tallow chandlers from storing tallow in residences. In the English-speaking world, this was undoubtedly the first nationwide fire prevention legislation enacted.

During the seventeenth century, a fairly complete code of building regulations was placed in effect in London and elsewhere in England, but no ready means of enforcement was provided for over 100 years. After this time, progress in the fire prevention field began to pick up speed.

Early America and Fire Safety

The first fire prevention ordinance in the New World was enacted in Boston in 1631 following a major fire that threatened the life of the new town. This ordinance prohibited thatched roofs and wooden chimneys and included provision for enforcement. Later Boston building codes required stone or brick walls and slate or tile roofs for many types of buildings.

During the early eighteenth century, mutual aid societies were developed in various American cities. The primary purpose of these societies was the salvaging of goods exposed to fire. While these societies were not engaged in fire prevention as generally understood today, they did seek to limit the fire damage potential by means other than fire fighting and to that extent were performing a fire prevention service.

About 1835, industrialists, particularly in the textile industries, acted on their concern over the number of large plants destroyed by fire, and formed insurance companies to combat the effects of such losses. Originally, these insurance companies primarily collected premiums and reimbursed plant owners for losses. However, while this was being done, farsighted individuals were studying fire losses in efforts to determine what caused fires, why small fires became large, and what practical measures could be taken to prevent fires and to limit the spread of fires when they did occur. This giant step forward provided the real base from which modern fire prevention has emerged.

Advances in the Nineteenth Century

During the latter half of the nineteenth century, industrialists constructed larger and larger buildings to meet the needs of rapidly developing technology. It soon became evident that the types of construction in general use were not adequate to meet the fire problem. Buildings with outer masonry walls but with **combustible roofs and** floors were seen to be subject to internal collapse and possible exterior wall failure in the event of a major fire.

At first, attention was focused on noncombustible construction without regard to occupancy in the belief that such construction was fireproof. Experience soon proved that exposed steel or cast iron framing was not fireproof, and considerable knowledge was gained on the yield points of steel and other structural materials under fire temperatures. (The yield point temperature for steel, for example, is well below temperatures in a major fire.)

Builders attempting to construct fireproof plants also experimented with various types of masonry arches supporting several types of floors. But experience showed that such construction might collapse from fire or to relieve a water load—obviously a costly field demonstration of what kind of arches, what kind of floors, and what kind of fillers are adequate to withstand a serious fire.

The closing years of the century saw the introduction and increasing use of the concrete and reinforced concrete construction that is so widely used today. As a result, much was learned about the fire capabilities of such construction—for example, the fire resistance of various types of concrete and the requirements for coverage of reinforcing rods.

Along with advances in knowledge of building construction came a recognition of the need for better means of controlling fires once they had started. The first step beyond hose lines, water buckets, and the like was the installation of perforated pipes. The second step—primitive open sprinklers—quickly followed. These systems worked to some extent, but by their very nature they caused more water damage than was necessary in most instances, and they were susceptible to clogging. A number of individuals, recognizing both the inherent deficiencies of these systems and the potential for vastly improved fire control, worked on ways to make them automatic. About 1878, the first practical automatic sprinkler system was installed. From then on, the development and installation of automatic sprinkler protection progressed rapidly, and fire losses dropped sharply.

At the same time that these advances in construction and protection were occurring, insurance companies and other interested organizations were making studies of the part that occupancy (the use to which a building is put) played in the fire loss picture. From these studies, it became possible to draw some conclusions as to ways existing occupancy practices could be revised to provide more security against fire without harm to operating productivity. From these conclusions came the first detailed fire prevention regulations.

It is proper to say that, during the nineteenth century, construction, protection, and occupancy were for the first time brought together to assure greater fire safety than had ever before been achieved. From this solid base, succeeding generations of men and women involved in fire safety have brought into being the present building and fire codes on which fire prevention programs now rest and from which these programs should move forward to ever greater effectiveness.

Fire Prevention and the Technological Revolution

When most people think of the technological revolution, they think primarily in terms of developments that have made life easier or more pleasant—for example, television, radio, air conditioning, power tools, and shorter hours of labor. Too few realize that many of these developments are also adaptable to fire safety. As a result, there have been many needless tragedies, of which the following are examples.

1. 1903, Iroquois Theater, Chicago, 602 dead. This tragedy was due to a combustible screen and other combustibles on stage, poorly arranged exits, lack of smoke vents, and other defects, all of which could have been eliminated with the technology then available.

2. 1911, Triangle Shirtwaist Factory, New York City, 146 dead. In this fire, most exits were locked and the keys were not readily available. There were communicating openings between some stories, and there was no effective automatic or first aid fire protection. Once again, a major fire tragedy could have been averted by the application of then known fire safety principles.

3. 1942, Cocoanut Grove, Boston, 492 dead. This nightclub had poorly arranged revolving doors, screened-over exits, limited emergency lighting, flammable decorations, and other deficiencies. Application of 1942 fire safety capabilities would have eliminated this disaster.

4. 1947, S.S. *Grandcamp*, Texas City, 468 dead, $67 million loss. Mishandling of ammonium nitrate fertilizer was responsible for this loss.

The significant fact about each one of these losses and about hundreds of other large fire losses of life and property is that almost all of them could have been prevented by the proper use of the technology available at the time of the loss. This is an area to which fire prevention personnel, or, for that matter, all those concerned, should address careful attention. It should not be necessary to have a disaster before corrective action is taken. Available fire safety technology should be used before, not after, the fire.

Some Errors in Current Fire Prevention Thinking

It is highly probable that the principal obstacle to improved fire safety for most communities and most organizations is the failure to recognize the need for better coordination among the various types of fire safety organizations and between these organizations and nonfire organizations that may be involved. It has been the author's experience that an outstanding fire prevention or fire safety record results when fire protection and prevention engineers or inspectors, fire department officers, fire marshals, and such nonfire personnel as building department and water department officials talk with each other on matters of mutual concern.

A second common error is thinking that fire prevention is limited to preventing the outbreak of fires. It is vitally important to develop means to prevent fire outbreak, but limiting the spread of fires that occur in spite of our best efforts is equally important. Limiting the spread of fire is not the sole responsibility of firefighting forces. The responsibility is shared by fire prevention personnel, who should recommend practicable modifications to building, occupancy, and protection so as to limit the spread of any fire that might occur.

Still another common error is the failure to think in terms of cost effectiveness. It does no good to recommend a solution to a fire problem if the solution costs more than the potential loss it is designed to prevent. All fire prevention personnel should be aware of the fact that, under most circumstances, it is possible to find an economically sound solution to fire problems found in industrial plants, mercantile establishments, places of public assembly, or elsewhere.

A fourth error is the failure to realize how far the shock waves from a major fire may spread. As an example of this the 1953 General Motors fire in Livonia, Michigan, might well be cited. This fire was responsible for six deaths and many millions of dollars worth of direct property loss. These losses, however, were only the beginning. First, operations at automobile assembly plants using the type of transmission manufactured at Livonia were of necessity curtailed. Second, operations at plants supplying either Livonia or the assembly plants were affected in varying degrees. Finally, thousands of employees at Livonia and elsewhere were thrown out of work. There is, of course, no way to accurately assess the cost of the indirect losses resulting from this fire, but it would not be out of line to assume that they exceeded the direct fire loss. Today, there are still many situations where similar large indirect fire losses could occur because of the failure to recognize the potential effects of the loss of a key plant or a key operation.

A Look into the Future. We cannot, of course, sweep the past fire record under the rug and forget it. The record is there, and we should use it as we move forward to better fire prevention programs in the future. But if any meaningful progress is to be accomplished, it is necessary that men and women in the fire prevention field develop certain attitudes of mind and be prepared to take some major steps forward.

Of primary importance is the need to reorient our thinking to the twenty-first century rather than stand still or look back to the nineteenth century. Such forward thinking is necessary because the technological revolution does not stand still. This century has already seen the advent of the automobile, the airplane, the submarine, spacecraft, computers, automation, tens of thousands of new materials and chemicals, open heart surgery, nuclear power, and more. All of these advances have been of great value in making life easier and better, but, at the same time, they have introduced fire problems our ancestors never knew. Along with the fire problems, however, new and far more effective tools for fire prevention and suppression have come into being. It is entirely in order to say that fire prevention has come of age.

As we look towards the twenty-first century, we should anticipate that the pace of technological change will continue to pick up speed. (Over 90 percent of all the scientists and engineers who ever lived are living now.) Fire prevention officials and organizations should prepare themselves to meet these changes. To do this, it is necessary to give recognition to certain facts as follows:

I. The fire prevention field today includes such diverse major subfields as construction, hydraulics, chemistry, nuclear physics, transportation, fire law, and electricity. It is manifestly impossible, for any one individual to become an expert in all of these subfields. Fire personnel should gain a general understanding of each of these subjects and, over a period of time, become expert in those areas with which they are directly concerned.

II. There are hundreds of organizations dealing with various phases of the fire prevention problem. Within these organizations is all the expertise necessary to find solutions to fire problems that are outside the range of the individual's competence. It pays to know that these organizations exist and what their capabilities are and to establish contact with them. A description of some of the major organizations is included in Chapter 2.

Fire Prevention Philosophy

The basic philosophy of all fire organizations and especially fire prevention activities should be one of service, dedicated to the protection of life and property against hazards from fire. Because of their service characteristics, fire organizations of all types have, in general, enjoyed widespread public support. Retaining this support, which is essential for true effectiveness, requires (1) being on the alert for ways and means to improve service, and (2) avoiding any activity that could jeopardize that support.

If we analyze existing fire prevention programs to find out why some are effective and some are not, we will find that failing or faltering programs have not adequately dealt with one or more of six problem areas, as follows:

I. Public apathy. How can public apathy be combatted if fire prevention promotional activities are limited to fire prevention week, a few lectures, and casual inspections of the larger industrial and mercantile operations? The answer is that these limited activities will not overcome public apathy. To arouse the public, it is necessary to develop year-round programs with special emphasis on personal contact with the public at large and on close, friendly relationships with the media.

II. Combustibility of environment. At the present time, few architectural or construction engineering courses provide any realistic instruction in fire safety. As a result, we have many instances of fire problems created by the indiscriminate use of wood shingles, the misuse of certain potentially toxic plastics, the installation of substandard electrical equipment, and other errors. Pending the provision of mandatory fire safety courses for architects and engineers, it is essential that every opportunity to get the fire prevention message across to these professionals be taken.

III. Proliferating ignition sources. As the technological revolution continues, there will be a continual increase in the number and variety of ignition sources. These ignition sources need not be uncontrolled or defective. Education of the responsible individuals is necessary.

IV. Building deficiencies. In all parts of the country, far too many buildings are being constructed without adequate exits, fire cutoffs, space separation between buildings, fire-resistive interior finish, and

History and Philosophy of Fire Prevention

the like. Appropriate building and fire codes plus provision for plan review by persons trained in fire safety are in order.

V. Private fire protection inadequacies. Far too often, not enough attention is paid to the adequacy of installed fire protection systems. Poorly designed or maintained systems are of little value in limiting the spread of fire. In addition, much worthwhile protection is lost because of the failure to understand trade-offs. For example, the ABC Co. proposes the erection of a five-story, fire resistive building with an occupancy that is in part combustible. A $50,000 sprinkler system is beyond the company's financial capacity. In this situation, a fire prevention representative could agree to enlargement of fire areas, slight reductions in fire resistive ratings, and other compromises that would reduce costs enough to permit the installation of the vitally needed sprinkler system.

VI. Research deficiencies. Too many products are coming on the market without adequate fire safety research. Use of such products should be tightly controlled. In addition, there is a need for the development of more accurate means for rating such characteristics as smoke emission and, in plastics, flame spread.

For a fire prevention organization to function at its best as a service organization, it is essential that problem areas be recognized and that appropriate steps be taken to solve the problems.

Even if the problems are recognized and solutions worked out, the entire advance can be jeopardized by improper presentation. Fire prevention organizations should never lose sight of the fact that they are service organizations and are there to advise and not command. When a situation requires police-type action, the matter should be turned over to the appropriate authority.

In summary, we might define fire prevention services as all measures taken to—
1. eliminate fire causes;
2. provide safeguards against inherent hazards that cannot be eliminated;
3. minimize the possibilities of small fires becoming large;
4. prevent fire deaths and serious injuries;
5. minimize fire's consequences for the business and social life of the affected community;
6. accomplish the above objectives at a reasonable cost.

chapter 2
Fire Prevention Organizations and Their Functions

During the past hundred and fifty years, a great deal has been learned about fire behavior and about ways of eliminating or at least minimizing the threat to life and property from wildfires. To the extent that this knowledge has been put to use, human life has become safer, the hazard to property been lessened, and the means of livelihood made more secure.

The development of modern fire prevention practices was begun primarily by private organizations set up to meet particular fire problems (for example, insurance companies covering the textile industry) and by local governmental agencies to meet peculiar individual needs. In general, the federal government was not involved except for federal agencies. As time went on, some of these original groups expanded to cover additional problem areas, and some groups became national in scope. The organizations' very nature, however, precluded them from covering the whole field of fire prevention. To some extent, this point of potential weakness has been overcome by cooperation among groups with various areas of expertise.

During the past twenty-five years, the federal government has become increasingly involved in fire prevention activities outside federal agencies. If, as seems likely, present trends continue, within a few years the federal government will be providing a clearinghouse for fire prevention information plus a training center for some types of fire prevention personnel.

It is estimated that between 50 and 75 percent of the things used by a forty-year-old adult have been developed since that person's birth. It is further estimated that 90 percent of the things used in the year 2000 by a person who reaches the age of twenty-one that year will have been developed since his birth. Given such a rate of technical progress, no one individual or organization can expect to become an expert in all phases of fire prevention. Rather, the aim should be to have a broad knowledge of the field, expertise in principal areas of concern, and knowledge of where to obtain expert guidance on special problems.

In addition to specific professional concerns, the fire prevention engineer, technician, or inspector should bear in mind at least the following four economic facts of life.

1. Any structure, vehicle, tool, or material to be usable must be able to perform its designed function. (For example, a house must provide shelter and a car must provide transportation.)

2. No structure, vehicle, tool, or material should present an unacceptable risk to life, property, or the ability to make a living. (This is the reason for codes.)

3. There is always an economic solution to the fire problem of any necessary operation that in its present form presents an unacceptable risk. (The solution should be practical and cost-effective; gold-plated recommendations are self-defeating.)

4. Buildings, vehicles, and tools can be designed and materials selected to meet both operating and fire safety reuirements.

Over the past thirty-plus years, the author has had many direct and indirect contacts with fire prevention personnel representing a number of organizations, both public and private. These contacts have been consistently mutually helpful. It is therefore recommended that persons planning a career in fire prevention make it a point to become acquainted with these sources of assistance and to learn the areas in which they function. The following list of organizations, which details some of their major functions, is given to provide a broad base on which contacts can be built.

Federal

I. National Fire Prevention and Control Administration. This federal agency, a new arrival on the scene, can perhaps be best described as an idea whose time has come. It represents the broad entry of the federal government into the fire prevention picture. How rapidly the agency will develop and what its ultimate mission will be, only time can tell, but at this writing it seems reasonable to predict that it will in the field of fire prevention eventually—

A. develop national fire prevention policies in those areas that are of national concern;

B. develop national standards that will draw heavily from standards developed by major private organizations;

C. provide a national clearinghouse for fire prevention information; and

D. provide facilities for training fire prevention personnel and assistance to nonfederal agencies also engaged in fire prevention training.

II. Department of Defense. All the major branches of the armed forces have strong fire prevention programs. All three of the major services maintain fire research facilities. All three have developed for the layman technical manuals on the proper testing and maintenance procedures for various types of fire protection equipment. In addition, they have developed regulations for insuring the safety of federal property. Most of the publications referred to are available to nonmilitary fire prevention personnel. Listed below are certain other special assistance capabilities that could at times be of vital importance to local fire prevention personnel.

A. Department of the Army. Highly trained explosives ordnance disposal units are stationed throughout the country. These units are readily available to assist local authorities in explosives problems.

B. Department of the Navy. A special fire prevention course, open to nonfederal employees, is regularly offered at the naval facility in Norfolk, Va.

C. Department of the Air Force. Several fire prevention schools are maintained. At times, these schools are open to nonmilitary personnel.

III. Department of Transportation (DOT). This agency has its own directly assigned fire prevention activities and also includes a number of agencies with an interest in fire prevention. DOT agencies prominent in fire prevention include but are not limited to the following:

A. Coast Guard. This organization has responsibility for the safety of waterfront facilities handling hazardous cargoes and vessels operating in American waters.

B. Federal Aviation Administration, Federal Highway Administration, and Federal Railroad Administration. These agencies are responsible respectively for the fire safety of air, road, and rail vehicles engaged in interstate commerce. The rail administration also oversees pipeline safety.

C. Hazardous Materials Regulations Board. This organization is responsible for the development of regulations governing the transportation of hazardous materials.

Copies of HMRB regulations and other DOT publications dealing with hazardous materials are available to personnel with a need to know. In addition, the DOT has developed a system of standard symbols to identify the nature of hazardous materials.

It seems appropriate to note here that private companies with fleets of rail tank cars have plainly marked the cars with company symbols, usually a combination of letters. These symbols can be very helpful in identifying a product in case of fire.

IV. Bureau of Mines. This subsection of the Department of the Interior is concerned with the fire safety of mines (coal, metallic, and nonmetallic). The bureau offers numerous publications on mine safety and, from time to time, conducts mine safety seminars that are open to the public.

V. National Bureau of Standards. This agency over a period of many years has done a great deal of research work in fire safety. For example, the bureau has investigated the fire behavior of building structural components, interior finishes, and many other types of materials. The agency's publications are available to the public.

VI. Forest Service. Principal agencies in this area are the U.S. Forest Service, in the Department of Agriculture, and the National Park Service, in the Department of the Interior. Both agencies have done considerable research in ways and means of limiting woodland fires. In addition, each has specific firefighting responsibilities. Publications of both agencies are available to the public.

VII. Occupational Safety and Health Administration (OSHA). This agency is charged with the responsibility of developing and enforcing appropriate fire safety regulations. These regulations apply to a large segment of private manufacturing, storage, commercial, and institutional activities as specified by Federal law. It has adopted with modifications a number of fire codes developed by the National Fire Protection Association, a private organization described below. These standards and other fire safety documents are readily available from OSHA. The standard guide to OSHA regulations can be obtained from the National Fire Protection Association.

State

Within state and regional governments there are many types of organizations dealing with fire safety. Organization names and responsibilities differ considerably from state to state and from region to region. For the purposes of this text, therefore, only the three main types of organizations, together with their usual principal type of duties, will be described.

I. Inspection and rating bureaus. In most states, if not all, there is an office or agency that deals with the rating schedules filed by insurance companies. Many such agencies also exercise administrative authority over the state fire marshal. The agencies generally have no direct fire prevention responsibilities, but the administrative authority cited can affect the overall state fire prevention program.

II. Fire marshals' offices. In general, state fire marshals are responsible for—

A. investigation of fires of suspicious origin or in which there was loss of life or large property loss;
B. compiling fire loss statistics;
C. establishing and enforcing regulations governing the use and transportation of hazardous materials within the state;
D. development of other fire-related regulations or codes as appropriate;
E. serving as a clearinghouse for information on state codes and regulations relative to fire safety;
F. inspection of certain types of premises as specified by law; and
G. basic regulations on fire detection and extinguishing systems.

However, the various state fire marshals' offices differ in their duties and in their place in state government. Fire prevention personnel should make it a point to know conditions in the state or states in which they are working, because state fire marshals' offices can be very helpful.

III. Forestry services. State requirements for forestry services are by their very nature dictated by geographic and climatic conditions. The rain forests of the Olympic Peninsula, the plains of Kansas, and the forests of southern California obviously have very little in common with one another. Accordingly, each of the state forestry services is geared to the needs of the state served. Working in conjunction with the United States Forest Service, these agencies have accomplished a considerable amount of fire prevention through hazard control on lands they administer and through public education.

Municipal

Fire prevention activities in municipalities differ widely both in the ways in which they are organized and in their overall effectiveness. For example, a city may have a department of public safety, a separate fire prevention agency, or a fire prevention section of the fire department. Any of these organizations can function well and any can do an ineffective job, depending upon how well it is staffed, funded, coordinated with other agencies, and supported by both responsible authorities and the public.

Municipal fire prevention bureaus or personnel can and should keep the public informed about fire safety, provide educational services as required, make fire safety inspections at the local level, and coordinate their own activities with those of other departments whose actions can affect the overall fire safety picture. For example, it is far easier and much less expensive to correct a fire safety error in the blueprint stage of a building than it is to correct the same error after the building is erected. Similarly, if there is mutual confidence, it is possible to eliminate hazards to life and property at an industrial plant before a fire drives the lesson home.

Private

I. National Fire Protection Association (NFPA). Of all the private organizations engaged in fire prevention work, the nonprofit NFPA is undoubtedly the best known. Its basic objectives are to develop and circulate fire prevention and protection information and to secure the cooperation of its members and the public in establishing proper safeguards against fire hazards to life and property. To accomplish these objectives, the NFPA engages in many types of activities, of which some of the most important from the fire prevention standpoint are as follows:

   A. *National Fire Codes.* These codes or standards of recommended practice cover such fire problem areas as construction, flammable liquids, flammable gases, flammable solids, portable and installed fire protection devices, water supplies, vehicles, fire departments, and others. All of these standards are drawn up by committees of technical experts from all walks of life. Standards are regularly updated to keep pace with technical developments. Of the 225 current standards, probably the best known are the *National Electric Code* and the *Life Safety Code.*

   B. Other publications.

   1. *Fire Journal.* This bimonthly is published for NFPA members. It covers technical developments, major and/or instructive fires, and other subjects of interest.

   2. *Fire Technology.* This quarterly deals with fire protection engineering and allied fields.

   3. *Fire Command.* This publication is designed specifically to meet fire department needs.

4. Books. The NFPA publishes books dealing with many phases of fire safety. The best known and most widely used is the *Fire Protection Handbook*.
  5. Occupancy fire safety studies and reports. These publications cover many types of occupancies.
  6. Educational materials. These are of many types.
 C. Field service projects.
 D. Coordination with fire prevention and protection organizations in other nations.
II. Factory Mutual System (FM). This organization, one of the world's largest industrial and mercantile fire insurance companies, provides its policyholders with many services in addition to insurance. Some of these services are also available to the general public. The most important of the services are listed below.
 A. Engineering services. These include plan review for fire safety, guidance on special hazards, and regular inspections of most insured properties by trained engineers.
 B. District offices and resident engineers located in all major industrial centers.
 C. Laboratories. These organizations conduct major projects in basic and applied research. They test for acceptability such diverse items as sprinkler equipment, safety controls, and fire characteristics of building materials. Results are noted in the *FM Approval Guide,* which is published annually. In many instances, equipment or material approved by the FM is marked with the organization's seal, the letters *FM* in a diamond-shaped frame. In addition to approval testing, the laboratories participate in the development of new or advanced types of fire safety equipment.
 D. Publications. Two important books, *Handbook of Industrial Loss Prevention* and *Loss Prevention Data Books* (a set) are available to the public, as is a variety of educational and training aids.
III. Factory Insurance Association (FIA). This association provides insurance services to a large group of major industrial, mercantile, and institutional organizations. In addition to insurance coverage, FIA provides engineering inspection services, engineering analyses of manufacturing problems and construction plans, and proposals for various types of fire protection systems and devices. There is also a laboratory that tests fire equipment and takes on research problems presented by industry. The FIA educational program is extensive and is supported by numerous brochures and technical pamphlets.
IV. Underwriters' Laboratories, Inc. (UL). The basic purpose of UL

is to examine and test systems, devices, and materials to determine their relationship to life, fire, and casualty hazards. The organization maintains laboratories in a number of states and foreign countries. Products meeting recognized national or international standards are listed in a series of booklets that form the UL Listings and are updated periodically. Other UL activities include but are not limited to the following:

   A. Development of new and improved testing procedures.
   B. Publication of fire safety brochures and other educational materials.
   C. Factory inspections of many listed products.

Many UL listed items bear the organization's seal, the letters *UL* within an oval. This seal signifies that the product meets the fire requirements for its designated use.

V. American Gas Association (AGA). The association operates laboratories in Cleveland and Los Angeles to certify that gas appliances meet recognized national standards for safety, efficiency, and durability. Appliances that meet test requirements are listed in the *AGA Directory of Certified Gas Appliances and Listed Accessories.*

VI. American Insurance Association (AIA). This nonprofit organization is probably best known for its publication of the *National Building Code* and a model *Fire Prevention Code*. These codes, which are periodically updated, are used in more jurisdictions of the United States than any of the other codes in widespread use. The AIA also publishes special interest bulletins.

Until fairly recent years the *Insurance Grading Schedule* (now known as the *Standard Schedule for Grading Cities and Towns*) was developed and implemented by the AIA. Currently the grading is being done by the Insurance Services Offices (an amalgamation of the property insurance rating organizations). Grading schedules are used to rate a municipality's fire defenses. Qualified engineers visit each municipality and grade its fire prevention services, building and water departments, fire departments, fire alarm systems, and other factors. Ratings and the recommendations accompanying them are used by insurance companies in computing insurance rates and by the municipal authorities to pinpoint and correct deficiencies.

VII. Bureau of Explosives. This arm of the Association of American Railroads investigates railroad fires and explosions and gathers information on the transportation of explosives and other dangerous materials. The bureau works closely with the Department of Transportation in the development of regulations designed to assure safe shipment. The Bureau of Explosives also maintains an inspection service and a testing laboratory and publishes special interest bulletins dealing with the handling of explosives.

VIII. Industrial fire prevention departments. Many large companies as a matter of policy name a ranking executive as responsible for fire safety. This individual generally supervises the company's fire prevention organization, large or small. In single plants, fire prevention may involve only one or two individuals; multiplant organizations may have fire prevention personnel at both national and local levels. The personnel may differ widely in their capabilities and training, depending on their responsibilities and on the nature of the plant or company operations. However, these individuals or organizations generally have certain things in common.

A. They have, or should have, a better acquaintance with plant fire hazards and the means available to combat them than their fellow employees do.

B. They maintain lines of communication with their fellow employees, who in turn may bring to attention hazards that might otherwise have been missed.

C. They have direct access to engineers from fire insurance carriers and can learn much from them.

D. They have, or should have, contact with fire department officers as appropriate.

With these and similar contacts, plant fire prevention personnel can accomplish a great deal in assuring plant fire safety with the backing of management. (They almost certainly have such backing or there would be no fire prevention personnel to begin with.)

Summation

The above list of fire prevention organizations is by no means exhaustive. An individual or organization with fire safety responsibilities can also obtain guidance or assistance from many other organizations, depending on local conditions. The author recommends that fire personnel familiarize themselves with all the fire prevention organizations appropriate to their own situation.

chapter 3
Building Construction

Far too often, reports of major fires (those involving loss of life, heavy property damage, and/or business interruptions) include clear evidence that defects in original construction or in later additions were major contributing factors to the extent of the losses to life and property. Such situations can be minimized if fire prevention personnel have a broad, not necessarily engineering, understanding of what fire safety in construction means and how to obtain the desired goals.

Fire prevention in buildings starts at the drawing board, where fire safety errors in original design can be corrected much more easily and at far less cost than would be the case with after-the-fact corrective action. The second line of action is to make sure the actual construction complies with the approved design. Finally, completed buildings of importance should be checked regularly for significant changes in construction or occupancy. If such alterations have taken place, the inspector should recommend construction changes for fire safety as necessary.

To accomplish the desired fire safety goals, fire prevention personnel should have a working knowledge of the tools available to achieve the desired ends, understand how to use these tools, and understand their limitations. Because it is neither possible nor practicable for any set of guidelines to be complete, it is also important to have a general knowledge of those areas where defects are frequently found and to be alert to new developments in the construction field.

The question of insurance premiums should be mentioned at this point. Fire insurance policies normally cover direct fire losses to property and may or may not cover other property-related losses. The policies do not cover loss of life or serious physical injury. Accordingly, it is entirely possible that changes recommended for the purpose of improving property fire safety and simultaneously reducing insurance premiums may not be sufficient to eliminate undesirable risks to life. While it is true that in most instances construction that is fire safe for property is also fire safe for people, it is not always true. This fact should be borne in mind by all fire prevention personnel who have life safety responsibilities.

Building Codes and Ordinances and Their Uses

Building codes and the ordinances that provide for their administration and enforcement can play a vital role in fire safety if they are properly used and their limitations are understood. Codes normally cover basic structural elements, life safety requirements, building utility systems, fire cutoffs, exposure protection, interior finish, and other fire safety features. The codes are not, however, designed to cover every conceivable condition, nor would it be practical for them to attempt to do so.

Building codes and their counterparts, the various fire codes, have come into being primarily for three reasons: (1) to prevent recurrences of past disasters, (2) to provide means for the elimination and/or control of known present fire hazards, and (3) to anticipate to the extent possible the construction developments of the future so that appropriate fire safety tools will be available as needed. In all three areas, it is essential that habits of sound judgment be acquired and used.

To a considerable extent, many of the provisions now in building codes came about because people, at least some of them, learned from past disasters such as those at Chicago's Iroquois Theater (602 dead) and McCormick Place exposition center (multimillion dollar loss) and Boston's Cocoanut Grove nightclub (492 dead). From these and other disasters have come many of our present code requirements for exit facilities, theater stages, ventilation, fire cutoffs, fire aisles, and protection systems. After-the-fact prevention measures should not be dismissed casually. Rather, we should start asking ourselves how we might improve our approach to the fire problem so that fire safety measures that become so plainly evident after the fire might instead be taken before the disaster occurs.

Building Construction

Current building and fire codes, in general, are excellent tools for dealing with ways and means of eliminating or controlling known present hazards. To some extent, the codes also help to anticipate future problem areas. However, codes are of necessity lengthy documents, and revising them to cover changed conditions requires considerable time. This fact must be borne in mind when one evaluates a new type of structure, new materials, or other innovations that come into being after a code has been adopted.

In summary, codes are excellent tools but, like all tools, they must be used properly and with good judgment. Good judgment includes the recognition of factors that are not covered by existing codes.

Ordinances, as generally understood, are local devices used to set up local standards, to adopt other existing standards by reference, and to establish local enforcement procedures. There are many ways of handling enforcement, but in the author's opinion the "carrot and stick" approach is usually the most effective.

Code "Blind Spots"

Building and fire codes as we know them are a fairly recent development and represent a tremendous amount of work by many very able individuals. The ground to be covered, however, was so great that inevitably some areas of potential importance were omitted. Eventually these areas will be included in official documents, but for the time being it is essential that the fire prevention practitioner or student be aware that such "blind spots" exist.

I. Perhaps the most important "blind spot" concerns structural life span. For all practical purposes, no building code covers this subject to any important degree. The fact remains, however, that a high percentage of major fires occurs in older buildings that have features not acceptable under present codes. In addition, it is known that, over a long period of time, excessive exposure to unusually high or low humidity, the presence of some chemicals, and certain other occupancy conditions affect the fire resistance and flame spread ratings as well as the structural strength of a building. Pending code lifespan requirements, such conditions should be considered in evaluating a building.

II. Many, but not all, codes impose no limits on the area or height of Class A fire resistive buildings. Such restrictions as are imposed are fire code requirements for certain specific hazardous occupancies. In evaluating such buildings, we need to bear in mind the fact that all combustible contents can burn. Our fire resistive building may be intact after a fire, but unlimited areas can mean unlimited fire losses and heavy life losses. Where unlimited areas are permitted, control of the continuity of combustibles is essential.

III. Pressurized escape routes are another recent innovation that has yet to be covered by most codes. When pressurized corridors or stair towers are being considered, the possibilities for life or death need to be recognized. Assume an apartment house with a corridor under slight positive pressure leading to a fire tower that has slightly more pressure than the corridor. Doors to fire towers normally swing into the tower. Therefore, the positive pressure in the tower tends to act against the person trying to open the door. Unless it is possible to maintain very low positive pressure (1 to 3 oz. per sq. in.) it is essential that such pressurized stairways be provided with positive mechanical means for counterbalancing the pressure.

IV. Other subjects that are not covered in any detail by codes include, but are not limited to, climate, high-rise buildings, automation, air-supported buildings, and structures spanning railroads or highways.

None of these comments on code "blind spots" is intended to be critical of code writers, who have developed extremely useful documents. The comments are made to emphasize the fact that the pace of technological change requires constant attention to new developments.

Major Types of Construction, Old and New

I. Classes of buildings. Nearly all building and fire codes subdivide the various types of buildings into seven major classes, as follows:
  A. Fire resistive, Type A
  B. Fire resistive, Type B
  C. Protected noncombustible
  D. Unprotected noncombustible
  E. Heavy timber
  F. Ordinary
  G. Wood frame

For details on the requirements for each class—for example, fire resistive requirements, permissible flame spread ratings, necessary installed protection, and building utility systems—the reader is referred to the standard building codes or to one of the numerous excellent construction handbooks. Our concern at this point is to develop an awareness of potential structural defects or deficiencies that by their very nature cannot be covered in codes.

II. Exceptions to codes. Many buildings in use today, including many fire resistive structures, were erected before there were any detailed codes. Such buildings may have serious defects. Other buildings, constructed according to code requirements, have been altered so much that they no longer resemble their original class. Still others, also constructed according to code, have been combined with different types of structures or were originally of mixed construction. Finally, there are the newer types of buildings that no code fits. The fire prevention inspector, therefore, while using the appropriate codes to the extent possible, should always be alert for structural weaknesses that could contribute to the spread of fire or to unnecessary damage from heat, smoke, water, toxic gases, or explosive pressures. He should never forget that fire prevention includes limiting fire damages as well as eliminating fire causes.

III. Older buildings

A. The two most common problem areas in the older types of fire resistive construction are the use of cast iron and the use of brick or tile flat arches to support floors and roofs.

1. Cast iron. Many old buildings have frameworks of cast iron columns, beams, and girders, most of which are exposed or lightly protected. Cast iron is brittle and may break from a relatively moderate impact or from a rapid temperature rise. Cast iron also has a relatively low tensile strength, and, in addition, old castings do not always have a uniform thickness of shell. Under these conditions, if a relatively minor fire in an upper story caused a load collapse, structural failure of a nonfire floor could easily result. This does not mean that these buildings should not be used. It does mean that there is a requirement for judgment in the selection of occupancies and for consideration of fireproofing of exposed cast iron.

2. Brick or tile arches. There are still a number of old buildings with essentially flat or slightly rounded brick or tile arches supporting roofs and floors. In industrial buildings, outward thrust of the arches is taken up by tension tie rods extending between steel beams or, in a few instances, between exterior walls. In many buildings the tie rods are exposed, but in others they are concealed above false ceilings of limited fire resistance. Where such ceilings exist, given the lack of fire knowledge at the time, it should be assumed that the space above is continuous and that it contains combustibles. (See Figure 1.) In reviewing the situation at a building with this type of arch, some of the factors that need to be considered are as follows:

   a. If sprinkler protection is added, do not cut into the arch to install hangers without knowing the type of tile or brick or

Fig. 1. A flat masonry arch

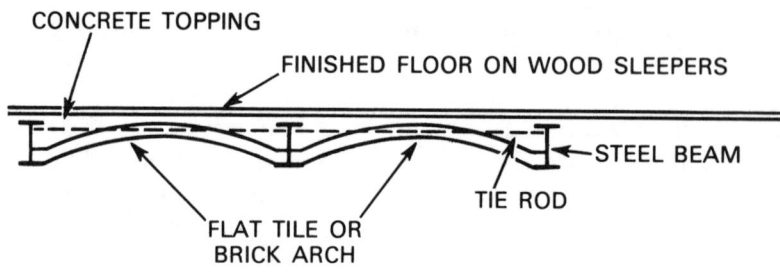

without providing for possible arch reinforcement from below.

b. Tie rods yield when exposed to high heat. Unless supporting walls are very heavy, the outward thrust of the arch may then collapse the walls.

c. If there is a possibility of heavy water leakage from above, it may be necessary to waterproof the floor above and provide drainage other than through the arch. Cutting an arch to release water can be extremely dangerous.

B. Balloon frame buildings. Another type of old-style construction that is still with us is balloon frame. (See Figure 2.) In such buildings, single or spliced studs run directly from the foundation through two or more stories to the eave line. At each floor level a ribbonboard is nailed to the studs. Floor and ceiling joists are nailed to the ribbonboards, and the floors and walls are attached to the studs and joists without any firestops. As a result, joist and stud channels open directly into each other, and fire once in the wall can spread quickly throughout the building. For such structures, the only feasible fire prevention measures, assuming the building is worth it, are to have noncombustible insulation blown into the concealed spaces and perhaps to install sprinklers. It should be noted that sprinklers alone are useless because they cannot reach a fire inside the wall.

IV. Modern construction. In modern construction, good judgment in the selection of buildings is vital. Three illustrative examples follow.

A. The ABC Co. manufactures cocoa and desires a Class A building for its grinding operation. Two buildings are available; one has flat reinforced concrete floor slabs on mushroom columns, and the other has pan-type concrete floors on beams on girders. In this case, the

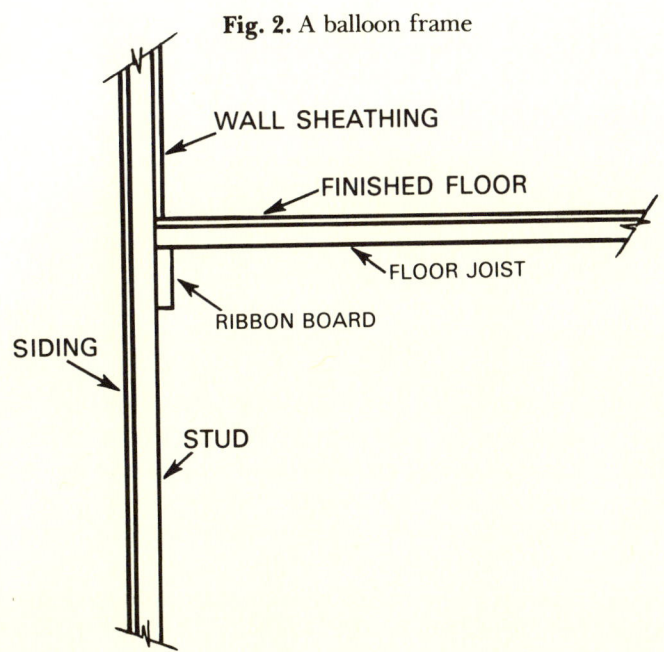

Fig. 2. A balloon frame

logical choice is flat slab on mushroom columns because, with such construction, the combustible dust control problem is virtually eliminated as a hazard.

B. The DK Co. has products that are highly susceptible to water damage and the story above has a high-hazard occupancy with the probability of frequent small fires that trigger sprinkler water discharge. If the floor above is straight reinforced concrete with a granolithic finish, there is little likelihood of serious water leakage. On the other hand, a concrete floor with cinder fill leaks readily and may leak for days. The author knows of a number of instances where water has leaked through a cinder concrete floor in minutes and then gone on to leak through additional similar floors below.

C. Not too long ago, the author visited a major building that had originally been constructed as a Class A building with heavy masonry walls and solid concrete floors. Areas were largely open. Today, each story has been cut up into small units separated by wood stud partitions lined with plywood on both sides. Hung ceilings of combustible fiberboard on wood frame were also installed. For all practical purposes, that building has been converted to a wood frame structure inside a Class A building.

Many more illustrations could be given, but these should be sufficient for our purpose, which is to emphasize the need to really see what we look at and the need to come up with practical solutions to the problems.

Hazards for Building Under Construction or Undergoing Major Alternations

Buildings under construction or undergoing major alterations are inherently more hazardous, regardless of the type of construction, than finished structures. Primarily, this is because a building under construction has less effective means to limit or extinguish fires than a building generally has.

Construction operations can be greatly improved from the standpoint of both fire prevention and fire protection by advance planning and by establishing project responsibility. The basic considerations involved are site preparation, control of temporary buildings, fire protection, and control of inherent construction hazards.

I. Site preparation. Most builders clear a site of vegetation and debris prior to the start of construction, but many make no provision for maintaining adequate access or for the installation of essential water supplies and hydrants. The author knows of two recent instances where large apartment buildings under construction were destroyed because responding fire departments could not get within 800 feet of the buildings and the water mains were not yet in service.

II. Temporary buildings. Temporary buildings to house offices, shops, and supplies are normally present at any large construction site. Far too often, they are grouped together so that a fire in one can easily result in the destruction of all or most of the others. In addition, too many temporary buildings are located inside of or too close to the building under construction and form a threat to it. Some of the typical hazards found in temporary buildings are as follows:

A. Defective heating appliances.
B. Poorly installed temporary wiring.
C. Inadequate ventilation, particularly in paint and engine shops.
D. Lack of any fire protection.
E. Careless handling of flammables.

III. Fire protection. Whenever a building of any size is being erected, substantial amounts of combustible building materials and highly vulnerable scaffolding and hoists inevitably are present during most of the construction period. Since this situation is unavoidable, fire prevention planning should provide for limiting the amount of combustibles stored and in use to the minimum practicable, using non-

Building Construction

combustible scaffolding and hoists where possible, and installing and placing in service fire protection systems as rapidly as construction progress permits. Standpipes and sprinkler systems should be in service not more than two floors below the highest construction.

Construction of any multistory building inevitably involves unprotected floor openings. Such openings should be closed up and stairs, elevator shafts, and pipe chases enclosed as rapidly as construction permits.

IV. Control of hazardous operations. A number of potentially hazardous operations are conducted at any construction site. Some of the most common are as follows:

A. Cutting and welding. Because of the inevitable presence of floor and wall openings, combustible materials, and combustible debris, it is essential that these operations be tightly controlled to minimize the possibility of fire from sparks.

B. Flammable liquids. Proper storage, dispensing, and usage of flammable liquids should be required. Particular attention should be paid to safety cans and their condition and to the disposal of paint-soaked rags. If at all possible, no more than one day's supply of flammable liquids should be stored in the building.

C. Trash disposal. Scrap combustibles and other construction debris should be removed from the premises daily. If scrap is to be burned, the site selected should be outside the building and away from combustible materials.

D. Tar kettles and roofing mops. Use of tar kettles involves the burning of fuel, the heating of a combustible material, and the possible exposure of combustible construction. Kettles should be located outside of buildings and should never be placed on a wood roof. Roofing mops, once soaked with tar, are susceptible to spontaneous ignition and should never be left inside a building or near combustible materials.

E. Temporary wiring. Assuming a proper installation by a competent electrician, the principal hazards are mechanical injury from construction equipment and the introduction of an operation that produces flammable vapor.

F. Gasoline-powered equipment. Engines for such fixed equipment as hoists, pumps, and compressors should be located so that exhausts are well away from combustibles. Mobile equipment such as lift trucks should always be refuelled and preferably be stored outside the building.

V. Control of alterations hazards. Hazards during major alterations are similar to those occurring at construction sites, and precautions should, in general, be along the same lines. However, because of the fact that the building being altered may be occupied, there are several additional subjects that need consideration.

A. If the building is occupied, planning must provide for maintaining adequate means of egress for occupants.

B. Existing installed protection systems that may be involved in the alterations program should be maintained in service to the maximum extent possible.

C. If removal of permanent fire barriers is necessary, substantial temporary enclosures for wall and floor openings should be provided promptly.

Each building under construction or undergoing major alterations presents a unique problem to which there may be no single fire prevention answer. It is the inspector's responsibility to apply basic guidelines for primary areas of concern and, through careful observation and good judgment, develop fire prevention programs that will minimize the fire risk to the buildings under study.

Horizontal and Vertical Fire Cutoffs

Building codes and fire codes both include many references to fire walls, fire partitions, stair and elevator enclosures, pipe chases, floor openings, and situations where fire cutoffs are or should be provided. An inspector should, of course, be familiar with the general requirements for horizontal and vertical fire cutoffs, but that in itself is not sufficient. He should also, bearing in mind that limiting fire spread is a major part of fire prevention, learn to judge whether or not the cutoffs as installed, maintained, and used are capable of performing their intended function.

I. Fire shutters. Assume for our first case that a factory has installed fire shutters at window openings to protect against a serious exterior exposure and that the shutters are closed when the plant is not operating. This is fine if there is an exposure fire, but what about an interior fire in a locked plant? Normally, very little attention is paid to this possibility nor to the desirability of having from one-third to one-half of the shutters operable from outside by responding fire organizations.

II. Rolling steel fire doors.

A. Location. Rolling steel fire doors are excellent fire cutoffs in the right places; unfortunately, they are often installed in the wrong places, as listed below.

B. Exitways. Rolling steel fire doors should never be installed in a

Building Construction

principal means of egress. When these doors operate, they can fall fast enough to seriously injure anyone trapped in the opening. Secondly, opening a closed rolling steel fire door is a job for a mechanic. Untrained personnel in all likelihood would be unable to open the door and could be trapped on the wrong side of the door.

C. Passageways for stock. Unless a mechanic capable of resetting rolling steel doors is available, operation of a door because of fire or another reason can shut down plant operations. The author is familiar with a plant where accidental tripping of a rolling steel fire door at a conveyor opening shut down major plant operations for twelve hours.

III. Sliding fire doors. Considering the simplicity of their operation, sliding fire doors in theory should present no serious problem. In practice, there are several problem areas, the neglect of which has been responsible for the spread of numerous fires far beyond their area of origin. For example:

A. Track lubrication. Unless the track is lubricated and kept rust free, the door will fail to operate in an emergency, or it will operate so slowly the fire will be through the opening before the door closes.

B. Chain length. Chains provided for counterweighted doors stretch. If the chains become too long and are not shortened as appropriate, the counterweights will be on the floor before the door completely closes. Fire will, of course, pass through the opening that remains. As an example of this, four sections of a California warehouse were recently destroyed because overlong chains kept four fire doors from completely closing.

C. Fusible links. That the operation of fusible links may be delayed by being painted is obvious, but too little attention is paid to the location of the links. They must be in the path of heat travel to be effective. If they are not, fire can pass through the opening before they operate.

IV. Fire curtains. These devices, while not a complete stop, are effective in high stories. Properly installed and maintained, they bank the heat and lessen the number of sprinkler heads that operate during a fire. As a result, the spread of fire at the ceiling is limited and the amount of potential water damage is reduced.

V. Escalators. Originally, escalators were installed in enclosed shafts similar to stair towers. Today's merchandising methods place the escalators out in the open so that customers riding them can get a good view of the merchandise on each floor. Several means have been developed to protect floor openings created by open escalators. Most establishments employ a combination of two or more of the principal methods tested below.

  A. Self-closing, rolling steel shutters, activated by fusible links or some other device, that move slowly so that no one on the escalator will be injured.

  B. Draft curtains and a line of close-spaced automatic sprinklers around the openings.

  C. Draft curtains with a line of closely spaced open sprinklers triggered by a pilot automatic sprinkler.

  D. Reversible ventilation triggered by a pilot sprinkler head or another device. In this system, fresh air is blown down through the opening while smoke-laden air is exhausted through a safety ventilation system.

VI. Pipe chases. Not too long ago, a minor fire extended through several stories of a pipe chase in the Empire State Building. The reason for the extension was simple. Workmen removed fire stops at various floor levels to expedite work and failed to replace them when the work was completed.

VII. Conveyor openings. These openings may be protected by various means, including miscellaneous types of fire doors and water sprays. For present purposes, we will limit comments to one aspect of water spray protection. In the author's opinion, water spray should never be used alone when there is only one water supply. Under such a condition, failure of the water supply leaves the plant with no protection for adjoining fire areas.

VIII. Expansion joints. When occupancy conditions permit, it is desirable to have a self-supporting fire wall at a building expansion joint (assuming a wall is needed). If this is properly done, there will in effect be an independent structure on each side of the wall.

IX. General recommendations. Many a fire cutoff deficiency has remained uncorrected and many a legitimate recommendation has been questioned because of faulty judgment on the part of the inspector. Two typical examples of faulty judgment are as follows:

  A. A four-story joisted building has open stairways. Fire resistance of the floors is thirty minutes. Going strictly by the book, the inspector recommends that the stairways have enclosures of concrete block with a fire resistance rating of two hours and with automatic-closing Class B doors at openings. No sale, and no wonder! It is pointless to recommend a stair tower that will stand by itself long after the building has been destroyed. It would have been far better to recommend a stairway enclosure of gypsum board on wood frame

Building Construction 33

and solid core wood doors, both of which have approximately an hour's fire resistance. The desired end would have been achieved at less than half the cost.

B. Thirty-odd years ago, the author visited a brewery that had just acquired a group of small, one-story, brick-joisted buildings separated from one another by fire walls. The total area was under 40,000 sq. ft., and the buildings were used for the storage of glass bottles in wood crates. The buildings were sprinklered. Conveyor openings were cut in the fire walls, and a previous inspector had recommended fire doors at each of the openings at a cost of several thousand dollars. Since none of the fire walls was needed in this small, protected structure with ordinary hazard occupancy, the author "killed" the recommendation. Always make sure that any recommendations that are made are really necessary and that they are cost-effective. Recommendations should not cost more than the losses they are designed to prevent.

Concealed Spaces

The building construction with which most of us are familiar has concealed areas in such places as cavity walls, stud channels, joist channels, and pipe chases. Normally, such areas are relatively small, and the fact that they may contain electrical wiring, gas piping, or plumbing is well known. In these areas, we are primarily concerned with the condition of the wiring, the provision of ventilation if gas may enter the space, the presence or absence of insulation, whether or not the back sides of wall and ceiling linings are combustible, and whether or not the original fire stops are still there.

In recent years, both business and industrial buildings have increasingly included a wide variety of other types of concealed spaces. Some are created by hung or dropped ceilings, furred out walls, or soffits; others consist of concealed rooms or sealed crawl spaces. Most such spaces are installed to conceal utility or process piping, ducts, or equipment; to hide work areas from public view; or for decorative purposes. These areas differ from the older types of concealed spaces in that their volume may be much greater and they may contain many different types of hazards. An inspection of concealed spaces should especially include the features listed below.

I. Materials. The materials used for ceilings, furred out walls, and soffits may range from fire resistant vermiculite and plaster on wire mesh to sheet metal to low-density fiberboard. Materials for concealing rooms or sealing crawl spaces are usually gypsum board on wood or metal studs or some form of masonry. Ceilings or furred out walls may be perforated or have a smooth surface. From the standpoint of materials alone, there are several fire problem areas that require attention, as follows:

    A. Combustible ceilings. The inspector needs to judge how large an area can be permitted, regardless of occupancy, without automatic sprinkler protection.

    B. Presence of flammable oils, chemicals, or vapors. Where industrial processes produce air-borne flammables, periodic checks should be made to ascertain that noncombustible ceilings and walls are still noncombustible. More than one building has been destroyed when a noncombustible wall or ceiling material, impregnated with flammable residues, became ignited.

II. Painting. Flat ceiling and wall materials are frequently painted. Several coats of oil-based or other combustible paint can make a noncombustible ceiling combustible.

III. Hanging and furring methods. Hung ceilings may be supported by wire hangers and tee iron or aluminum frames, wood furring on wood, bar joists to which are secured tie iron frames, or wood furring attached to nailing strips on structural steel. Furred out walls are usually supported by wood frame.

IV. Blind spaces. Wherever hung ceilings or furred out walls are installed, blind spaces are created. These spaces may or may not contain combustible materials or sources of ignition. In most instances, access panels or removable tiles enable one to observe blind spaces. Some of the materials and equipment that may be present in a form or amount that creates a potential fire problem are as follows:

    A. Supports. Where ceilings are supported by wood and areas are large, there can be a tremendous fire load in the concealed area.

    B. Wiring. Wiring may be in conduit, have noncombustible insulation, or be wrapped with varnished cambric. Wiring condition should be observed.

    C. Steam and hot water piping. Unless proper clearances between piping and wood framing have been provided, the wood will gradually carbonize and then ignite.

    D. Air conditioning, ventilating, and heating ducts. Unless ducts are provided with fire dampers as needed, they can carry fire into and out of the concealed space. Occasionally these ducts will be found to be wrapped with combustible materials.

    E. Gas piping. The primary hazard from gas piping is leaks. Where gas piping is present, care must be taken to assure that correct venti-

Building Construction 35

lation has been provided. Propane is heavier than air, but methane is lighter.

F. Protection systems. Many blind spaces house sprinkler piping with drop nipples feeding sprinkler heads below the ceiling. The author knows of some instances where contractors omitted needed heads in the concealed spaces and of one instance where considerable supply piping that should have been in the concealed space was missing.

V. Ceilings, general. Aside from the normal problems of blind spaces above ceilings, there are several other possibilities that should be given attention.

A. Multiple ceilings. It is not uncommon for a hung ceiling to be installed below one or more existing hung ceilings. When a survey for a sprinkler system was made at a large club near Seattle, five successive ceilings were found. Spaces between the ceilings ranged from 6 in. to 6 ft., and there was wiring in each blind space. Inside and outside dimensions should add up.

B. Holes in floors overhead. It is entirely possible for fire on one floor to drop through an open hole into the blind space or for combustible debris susceptible to spontaneous heating to enter the space.

C. Water. Water used on a fire and entering a concealed space has to come out somewhere. The inspector should be alert for potential places for leakage.

VI. Soffits. Many mercantile establishments have soffits over display counters. Most such soffits contain recessed lights and miscellaneous wiring. There may also be gas piping, ducts, and protection systems. Many soffits are not accessible except when partially removed for maintenance purposes. Every possible effort should be made to learn what is in the soffit and how adequate the maintenance is.

VII. Furred out walls and concealed rooms. The usual type of furred out wall extends only a few inches from the main wall, and the blind space generally contains only the furring and wiring. For these spaces, the primary concern should be the presence or absence of fire stops, particularly if the furring is wood. The worst potential fire problem exists when a wall conceals an occupied room. If inside and outside dimensions differ by several feet, the odds are heavy that some inside wall conceals a room. Along with dimensions, it pays to notice the locations of windows in the outside walls. These windows should normally be visible from inside. If they are not, look for a concealed room. The author has seen a location where inside dimensions were 10 ft. less than outside dimensions. On investigating, he found a door to a concealed room, said door being lined on the customer side with a rack of merchandise. Inside the concealed room, about thirty people were working under congested conditions. Windows for the room were of the security type, and security bars were fixed in concrete.

VIII. Crawl spaces. Many buildings have crawl spaces under the first floor. In most instances, the spaces are accessible and ventilated. In a few instances, crawl spaces have been found sealed off and with no or little ventilation. Access to these spaces is limited because of the difficulty of entry. If such spaces contain gas piping, access panels should be provided and periodic checks made for gas leakage. Some years ago, gas leakage occurred in a relatively inaccessible crawl space under an apartment house. The leak went undetected, and gas filled the crawl space and moved up a pipe shaft until it reached an ignition source. The resultant explosion destroyed the building. Fortunately, the explosion occurred when most of the occupants were out; so casualties were much lighter than they could have been.

IX. Summation. There are innumerable varieties of blind spaces, and the inspector must therefore approach each case individually. His judgment on fire safety requirements should be based on the nature of the materials enclosing the space, the contents of the space, and the size of the space.

Interior Finish

Most codes applied by fire prevention inspectors define interior finish as exposed interior surface that forms an integral part of the building or is attached thereto.

I. General requirements. Certain interior finish requirements are nearly universally recognized.

A. All codes that are generally accepted specify permissible flame spread ratings for exitways; places of public assembly; walls, if buildings are over a certain number of stories in height; exposed surfaces in unsprinklered buildings versus surfaces in sprinklered buildings, hospitals, and other public institutions; back surfaces of furred out walls and ceilings; and rooms of over a specified size.

B. Some codes also required adherence to specified rates of fuel contribution and smoke emission. Unfortunately, present smoke emission ratings are of limited value because some materials emit most of their smoke at the beginning or the end of the burning cycle.

II. Temporary finishes. Some codes deal with paint, wallpaper, plastic films, floor coverings, and the like; others do not. From a practical standpoint of fire prevention, it would appear logical to take an approach approximately midway between the two positions, for reasons outlined below.

A. Paint. One coat of paint has virtually no effect on the fire hazard. But multiple coats, particularly of oil-based paint or lacquer, could contribute to a major disaster. Such a situation in fact occurred in the Winecoff Hotel fire in Atlanta, Georgia, in 1946.

B. Carpeting. Rugs can be moved, but wall-to-wall carpeting or its equivalent is essentially fixed. Such carpeting may be wool with jute backing, cotton, or, more likely these days, synthetic fibers on jute or rubber backing. Some combinations are essentially noncombustible; at the other extreme, some burn very rapidly or produce toxic fumes. There is a case on record of an apartment house tenant being killed by toxic fumes from a slowly burning hallway carpet. Said carpet fire did not seriously scorch or stain the hallway walls. It is desirable that carpets in public places have flame spread ratings of 25 or less and little toxicity.

III. Permanent finishes. Until fairly recent times, almost all permanent interior surfaces consisted of plastered walls, solid wood sheathing, or various forms of masonry. Since the mid-1930s, however, a wide variety of new nonstructural building elements has been developed and accepted. Some of these elements have noncombustible or fire resistive characteristics, but more are combustible—some extremely so. It is an easy matter to check out the "safe" materials for desired flame spread and fire resistance capabilities, and so the fire inspector should be most concerned with the combustible types, some of which are discussed below.

A. Plywood. Plywood can, of course, be effectively fire retardant treated. However, the process substantially increases costs; so, in most instances, plywood used for interior finish is untreated. While other factors are involved, the fire inspector's primary concerns in rating plywood are the adhesives used, the thickness of the plies, and the extent of restraint.

1. The melting points of plywood adhesives range from 200° F. to well over 300° F. The melting temperature has a substantial effect on the length of time the plywood can be exposed to heat before it begins to delaminate.

2. Plies range in thickness from fairly thick to extremely thin. The former burn like wood sheathing of the same thickness, but the latter burn like paper once delamination begins.

3. Restraint. Restraint in the form of baseboards, ceiling moldings, and chair rails helps prevent delamination.

B. Low density fiberboard. Many millions of square feet of this material have been sold and installed. Much of this type of material ignites easily, spreads fire rapidly, and emits considerable smoke. Many organizations require that the fiberboard be covered with gypsum board where life safety is a major concern, as in barracks and dormitories. Fiberboard's economy, however, makes it generally acceptable in sprinkler protected buildings. It has one other characteristic worthy of note. Fire can travel long distances through it unobserved, simply by burrowing through the pores. Fire retardant treated fiberboard is also available.

C. Particle board and punched hardboard. These materials have essentially the same characteristics as fiberboard, except that fire does not burrow through them.

D. Plastic sheets. These materials are increasingly coming into use for wall and ceiling finishes. Many of them are classed as slow burning, noncombustible, or self-extinguishing on the basis of small-scale laboratory tests. Such descriptions should not be readily accepted without further proof. Preliminary large-scale tests by the Factory Mutual System clearly indicate that large amounts of some types of plastic lining and insultation burn far more rapidly in the vertical position than tunnel tests indicate to be the case. For this reason, special test procedures are being developed to further test plastics. There are several other factors, as follows, to be taken into consideration.

1. Life of fire retardant treatment. Most fire retardant plastics are basically combustible compounds to which a flame inhibitor has been added. There has been some evidence that the fire retardance may gradually be lost, particularly when the material is exposed to excessive heat for prolonged periods. Such evidence is not, however, conclusive.

2. Toxicity. Under fire conditions, some plastics emit highly toxic fumes and dense volumes of smoke. Care must therefore be taken in the selection of plastics to be used in places of assembly or exitways.

E. Summation. Interior finishes can present a tremendous variety of fire problems, depending on the type of material selected. Correctly used, such finishes can help protect both lives and property from fire.

Interpreting Fire Tests of Building Construction and Materials

Within the United States are located some of the world's finest laboratories for testing building construction and materials. But it is a fact of life that far too few people have any real understanding of test procedures or of how to interpret test results. This lack of understanding is not limited to the general public by any means; many fire organizations also do not have as much understanding of fire tests as would be desirable.

I. Dimensions. Probably the most common error in judgment is the assumption that a given type of roof, floor, or wall assembly, rated for "x" hours fire resistance, will have the same resistance regardless of how it is used. This is a fallacy that in extreme cases can lead to a false sense of security and result in serious dangers.

Assume that a plan is being reviewed and that the plan shows a fire wall of masonry construction "y" inches thick. Specifications with the plan require a four-hour wall. Fire test records show that a test wall of the same materials and thickness achieved a rating of four hours fire resistance. Does this by itself mean that the proposed wall meets specification requirements? By no means!

Fire resistance tests are based on standard-sized specimens of 100 sq. ft. or slightly larger, with a minimum dimension of 9 ft. for either length or height. In contrast, walls are designed to fulfill a wide variety of requirements for length and height, and these variations can seriously affect the wall's fire resistance unless compensating factors are introduced. As a simple illustration, assume three walls of identical construction but with different dimensions. To simplify calculations, we will disregard the limited restraints imposed by the bonding of the wall to ceiling and floor. This reduces the wall for calculating purposes to a beam between two columns. Each of these walls is now acted upon by a fire-induced pressure of 8 oz. per sq. in. (Convert uniform load to a point load to calculate bending moments.)

A. Wall no 1. Length 12 ft., height 9 ft. Wall area exposed to fire pressure equals 12 x 9 x 12 x 12 or 15,552 sq. in. At 8 oz. pressure per sq. in., the fire pressure load is 7,776 lb. Bending moment from the load is $PL/4$ or (7,776 x 12)/4, which equals 23,328 ft.-lb.

B. Wall no. 2. Length 20 ft., height 20 ft. Area exposed equals 20 x 20 x 144, or 57,600 sq. in. Fire load is 57,600 x 0.5 = 28,800 lb. Bending moment is (28,800 x 20)/4, or 144,000 ft.-lb.

C. Wall no. 3. Length between columns 300 ft., height 30 ft. Area is 300 x 30 x 144 or 1,296,000 sq. in. Fire pressure load is 648,000 lb. Bending moment is (648,000 x 300)/4, or 48,600,000 ft.-lb.

Masonry walls have good compressive strength but very little strength in tension. When exposed to excessive tension forces, such as those created by too high a bending moment, such walls will break. In the illustrations above, wall no. 1 was approximately the size of a standard test specimen. Wall no. 2 was somewhat larger than a test specimen but, given the effect of restraints at ceiling and floor and some tension-resisting capability, would probably provide the required fire resistance. Wall no. 3 under no conceivable circumstances could be considered as likely to provide the needed fire barrier. It would fail long before the desired fire resistance time was reached.

In general, therefore, fire tests for construction assemblies can be considered valid for average dimensions, but, when these dimensions are exceeded, reinforcement is required. For walls, a length of 25 ft. and a height of 25 ft. could be considered average. If walls exceed these dimensions, thought should be given to the need for additional columns, pilasters, or other means of reinforcement. The author remembers a warehouse where lengths of wall between columns were excessive, as was the free-standing height. During a fire, this so-called four-hour wall failed in less than two hours. Had there been more reinforcement, the wall would have stood.

II. Piercing. A second common error is to disregard the effects of piercing or partly piercing a wall or floor for ducts, piping, beam supports or other equipment. When a 12-in. wall is penetrated 4 in. to support an exposed steel beam or a wood joist from a fire resistance standpoint the wall in that area is 8 in. thick. If beams are installed end to end from opposite sides that part of the wall from a fire resistance standpoint is little better than 4 in. thick.

Examples could readily be developed to demonstrate in other ways that fire resistance ratings for construction assemblies basically apply only to field assemblies of the same type and installed in the same way. When there are deviations—and there will be—a judgment factor based on experience and common sense must be applied to decide how much deviation is permissible and when recommendations for changes should be made.

III. Flame spread ratings. A third type of error is to assume that all flame spread tests are alike or that a rating applies to a material regardless of how it is used. Both assumptions are false. In general building materials are rated by the tunnel test method. A few materials, such as carpeting, are also rated by other methods. When more than one test method has been used, the inspector should compare the ratings. They can be quite different. In a tunnel test, samples are installed horizontally in what is essentially a ceiling position. Most materials have a somewhat different rate of flame propagation when installed on a ceiling that when installed vertically, sloping, or on a floor; but the differences for most materials will not be sufficient to cause serious concern. Some materials, however, have a radically different flame spread rating in a vertical or sloping position than in a horizontal position—particularly over large areas. (Plastics are a good example of such materials.) Here again the inspector should develop good judgment in determining what is permissible and whether or not the potential hazard requires the installation of a fixed protection system.

IV. Summation. Fire resistance and flame spread ratings are very useful fire prevention tools, but it is necessary to learn how to use them.

Effects of Major Occupancy Changes on Building Fire Safety

Many an inspector visiting a building has found a major change in occupancy and has made a careful sudy of the new occupancy but failed to check on whether or not the building itself was suitable for the new use. The safety of life and property in any part of the building could well be affected by major occupancy changes in the same area or in stories above or below. For example:

I. A block of multistory cold storage warehouses is converted to a connected series of loft buildings. Cold storage buildings normally have few people in any areas other than loading docks. Loft buildings may have substantial numbers of people in all areas. Question: Are the life safety facilities provided for cold storage occupancy still adequate for the changed operations?

II. A brick-joisted building at a hospital, formerly used for miscellaneous storage under sprinkler protection, is converted to a nursery for newborn infants. Because these infants cannot turn over, the sprinkler system in that area must be removed. This development immediately introduces a requirement for changing ceiling and floor construction to noncombustible or fire resistive.

III. The fourth story of a building has its occupancy changed from general offices with steel furniture to hazardous operations susceptible to frequent fires and consequent water discharge from sprinklers. The third story is occupied by operations that are highly susceptible to water damage. In many instances the only solution is to waterproof the floor.

IV. A warehouse of moderate size has been used to house structural steel. Occupancy is changed to storage of baled cotton. At this point consideration should be given to installing smoke vents in the roof.

V. Until recent years high-rise buildings were constructed with glass windows that could be opened. Today many of these same buildings have replaced original windows with sealed windows of shatterproof plastic. How does such a change affect exit requirements?

VI. A building is used for operations involving heavier-than-air flammable liquids, and suitable low-point ventilation is provided. Activities change, and flammable gases that are lighter than air are used. How should a building be altered to substitute ceiling for floor ventilation or to permit both to function at the same time?

The above examples illustrate the problem. Every time there is a major occupancy change, it is necessary to consider the building as well as the occupancy. In the life safety area, the requirements for number of exits, travel distance to exits, and permissible types of exits are determined by the number of people involved, thir ages, and their physical and mental conditions. Insofar as property is concerned, there are three basic categories: light, ordinary, and extra hazard. Generally speaking, if the category remains the same, no major building changes will be required except for such items as conveyors and their openings. However unusual susceptibilities to smoke and water damage may necessitate changes. On the other hand if the category changes to a higher degree of hazard, there well may be a need to increase fire resistance of walls, ceilings, and floors and to subdivide areas.

Occupancies do affect the acceptability of building construction.

Special Life Safety Requirements for Nursing Homes, Hospitals, Schools, and Similar Institutions

There are millions of Americans against whom there are extremely heavy odds should a fire develop in the building in which they are living or are temporarily housed. These are the aged, the very young, and the physically or mentally handicapped. Fire reports show that fire kills or critically injures such persons in numbers far out of proportion to their percentage of the general population. No one knowing the facts can seriously doubt that institutions, in general, probably present the highest occupancy hazard from the standpoint of life safety.

While it is quite true that much can be done to make the occupancy

Building Construction

safer, life safety for the handicapped begins with the building. These individuals have needs over and above those of physically and mentally capable adults. Some examples of the special problems faced by these groups are as follows:

I. Many persons in hospitals are on such life-support systems as cardiac monitors and kidney dialysis machines. Some of these people cannot be moved and others cannot remain off a life support system very long. Fire resistive construction by itself is not enough. Consideration should be given to smoketight doors and to the proper arrangement, construction, and protection of exits from rooms housing life support systems and leading to secondary support systems.

II. In nursing homes it is likely that some patients are bedridden, others senile, and all feeble. Such individuals are highly susceptible to panic. To the extent possible, construction should be such that large scale evacuation should not be necessary. If at all possible, the building should include several areas separated by smoketight fire doors so that patients may be moved from a fire area to a safe area without leaving the building. Should complete evacuation be desirable, exit routes should be arranged so that there will be no sight of the fire to excite the patients.

III. Certain institutions care for the crippled who can move about in wheelchairs. Such persons would be far safer from fire if ramps were installed instead of stairs.

IV. Many day-care centers are converted residences of ordinary or wood frame construction. In these centers, very young children often take naps alone or with other tots. Such children are peculiarly vulnerable to fire, and consideration should be given to requiring special fire safe construction.

V. Some institutions care for the blind, who cannot see an exit sign, or the deaf, who cannot hear an alarm. Means must be provided for safe egress of such people.

In the several cases we have cited and in many more types that could be mentioned, there is one common factor: time delay. The elderly, the infirm, the very young cannot be evacuated as rapidly as the same number of able-bodied and alert adults or, for that matter, trained older children. This simple fact means that unless special measures are taken, a fire would expose the handicapped to far more smoke, toxic gases, and heat than the general public would suffer. Care in the selection of permissible interior finish will go a long way towards correcting this situation.

No two institutions are exactly alike, nor are their occupancies.

There is therefore no simple answer to the problem of fire safety for the handicapped. The inspector should review each situation on its own merits and endeavor to come up with practical recommendations for sound construction, proper exitways, control of smoke, and other safety measures as needed for the occupants involved.

High-Rise Buildings

One of the chief fire prevention problems in high-rise buildings is that of fire spread. When any analysis is made of this problem it is essential that a clear distinction be made between old-style and new-style high-rise buildings.

I. Old style skyscrapers, such as the Chrysler, Empire State, and Woolworth buildings, are of substantial masonry and concrete construction. Many of the interior walls and partitions are masonry, plastered on both sides. Generally, exterior windows are glass and operable from the inside. Masonry curtains between windows on successive stories are usually high enough to limit the likelihood of fire communicating up the side of the building via exterior walls. Under these conditions, and with such sprinkler protection as occupancies warrant, it is a relatively simple matter to provide for both life safety and protection of property. Some years ago, a B-25 bomber struck the Empire State Building in an upper story. There were some casualties, and there was considerable damage to the impact area, but the rest of the building was not seriously damaged. A similar incident with a modern high-rise building could conceivably result in nearly total building damage plus heavy casualties.

II. Modern high-rise buildings, sometimes called glass-walled buildings, undoubtedly have some advantages in costs and in the flexibility of occupancy arrangements. They do, however, also present some fire safety problems that are not present in the older types of skyscrapers. Some of those problem areas are outlined below.

A. Windows.

1. There is for all practical purposes a continuity between the windows on one story and those on the next. If combustibles are present in either the interior construction or the occupancy, fire can travel upward story by story unimpeded by any physical barrier. Such fires have already occurred, those in Sao Paulo, Brazil, being especially noteworthy.

2. In many modern high-rise buildings, windows are also essen-

tially continuous horizontally. With such an arrangement, it is entirely possible for fire to bypass supposed fire barriers between sections.

3. To conserve heat and to simplify air-handling problems, exterior windows are generally sheet glass or shatterproof plastic, and no provision is made for breaking them or opening them in an emergency. In this situation, consideration should be given to the combination of internal construction and air handling that is needed to protect occupants against the spread of toxic gases from a fire. How are such gases to be limited in their formation, and how are they to be vented?

B. Smoke towers. Many older skyscrapers have smoke towers that are an integral part of the building. Open-air balconies leading from major sections of each story provide access to the towers. Such an arrangement is entirely feasible: because of the lack of upward fire movement, not more than one or two balconies would be untenable even in a major fire. In modern high-rise buildings, on the other hand, the upward fire movement possibilities make such a method of access impracticable. Smoke towers in such buildings must be designed so that they are accessible without undue exposure to smoke.

C. Evacuation and areas of refuge. Because of the length of time required to evacuate a skyscraper, a plan for occupant fire safety generally must include both evacuation and movement to preassigned areas of refuge. In the older high-rises, this is not a difficult problem because very limited travel would place a solid barrier between the occupant and the fire. Many modern high-rise buildings have no such barriers except for stair shafts. In such a building, a fire might be plainly visible. To prevent panic, exitways should be unusually well laid out, used in combination with areas of refuge in designated stories, and coordinated with internal communication systems.

III. General safeguards. Major high-rise fires—are they necessary? The answer to this question is no, provided a little common sense is used.

A. Limit the use of combustible materials in building construction and in occupancy of areas near the windows. Doing so will minimize the likelihood of a fire jumping vertically story by story via the windows.

B. Compartmentalize each story by either noncombustible or fire resistive partitions, by wide aisles, or by strips of noncombustible occupancy.

C. Select interior finishes that, in addition to being noncombustible or slow burning, do not emit dangerously toxic fumes under fire conditions.

D. Provide necessary protection systems. No two high-rise buildings are alike, and each requires different solutions to the problems of fire prevention. A fire prevention officer must approach each case individually, remembering that the objective is to minimize the risk to life and property at an acceptable cost.

Unusual Types of Buildings

Technological advances and plain economics have combined in today's world to produce several new types of structures that, as far as the author is aware, are not covered in any building code to any extent.

I. Air-supported structures. These balloonlike structures are constructed of flexible fabrics supported and held in place by a small positive pressure generated by fans. A sudden deflation during a fire is not likely, but it is possible. It is therefore of vital importance to plan internal arrangements so that exitways will not be blocked in event of fire and deflation. At present, the most common uses of air-supported structures are probably at construction sites where the structures permit work to progress regardless of the weather, and for temporary places of public assembly. The use of such buildings for any public assembly purpose should be discouraged.

II. Buildings over highways and rail lines. During the last few years, construction sites in urban centers have become increasingly scarce and costly. In many cases, superhighways and railroads occupy strategic locations. Accordingly, it has become increasingly common to design buildings to span highways and/or rail lines. For this type of construction, the principal fire threat is from below. How does one prevent a fire from a vehicular accident from damaging the building spanning the highway or railway? To answer this question it is necessary to know what kind of traffic passes under the building. If the traffic is limited to automobiles and/or passenger trains, that is one problem. If trucks or trains carrying flammables in tank car quantities are involved, substantially different construction is required.

There are numerous other unconventional types of buildings such as launching towers and geodesic domes but such buildings are not very common at the present time and, because of their very highly specialized nature, they require detailed technical studies.

Summation

Determining fire safety in building construction depends perhaps 20 percent on technical skill and 80 percent on common sense. Much of an inspector's expertise depends on habits of sound judgment and on the ability to really see what he looks at.

chapter 4
Fire Exposure

Few people, even those engaged in fire prevention activities, take a good hard look at their surroundings prior to entering a building. But fire safety personnel should do so, for a building's external exposures can lead to very serious fire problems and may have a great deal to do with the safety of people and property inside the building. Once an inspector makes a realistic evaluation of external exposures, he can evaluate the adequacy of the building's fire defenses against such hazards. If defenses are found to be inadequate, he can recommend realistic fire defenses.

In buildings of more than one story or with more than one fire area, consideration must also be given to the internal fire exposure problem. When the building is wholly occupied by the owner or by a single tenant, this type of exposure normally gets a reasonably adequate review because of the overall nature of the inspection. However, when multiple tenants are involved, there is an unfortunate tendency to overlook the potential for direct or indirect damage to the inspected area from fire in another story or in another fire area.

Many lives have been lost and many properties destroyed or seriously damaged as a result of external or internal exposure fires. These losses are not all from fire alone; in many cases, water, smoke, and toxic gases are responsible for a large part of the total fire loss. It is therefore important that all the major potential sources of damage from exposure fires be considered in order to make certain that adequate defenses are provided for the exposed property.

External Exposures

I. Judging the danger. An inspector must estimate the severity of the exposure, the potential intensity and duration of an external exposure fire, and the potential for conflagration.

  A. Dangers from exposure. There are seven main ways in which an external exposure fire can affect exposed property.

    1. Radiation. Heat can move laterally in the form of radiation. (A few years ago, heat radiated from a fire in an unsprinklered rubber tire warehouse, crossed a broad street and operated a number of sprinkler heads in a building more than 125 ft. away. Water damage in the second building was extensive.)

    2. Convection. Heat from a burning building moves outward and upward by convection. (On many occasions, stories well above an exposing fire have been affected by convected heat.)

    3. Conduction. Heat may spread through such conductive materials as unprotected ducts and pipes. Where such materials pass through a fire barrier, heat from the original fire can be conducted through a blank wall and ignite combustibles on the other side of the wall.

    4. Physical transmission. Flying brands and burning embers transmitted by the wind can ignite combustible surfaces far beyond the range of heat transfer by convection, conduction, or radiation.

    5. Floor leakage. Water discharged during a fire can leak through floors that are not watertight and seriously harm lower occupancies susceptible to water damage. In addition, there is potential for major water-staining of ceilings, floors, and furnishings, buckling of parquet or wood-block floors, and other structural damage.

    6. Smoke damage. Smoke travels by convection. Smoke from an external fire may enter an exposed building through any available openings. Once smoke enters a building, it may spread through air-handling systems or by other means. Many occupancies are highly susceptible to smoke damage.

    7. Toxic gases. These gases travel the same paths as smoke does. The more deadly types of toxic gases may cause asphyxiation at a point remote from the fire of origin.

  B. Examples of exposure hazards. The value of checking exposures may be demonstrated by making an analysis of the potential effects of an exposure fire on each side of a building in a typical city block

Fire Exposure 51

Fig. 3. Building exposures, plan view (no scale)

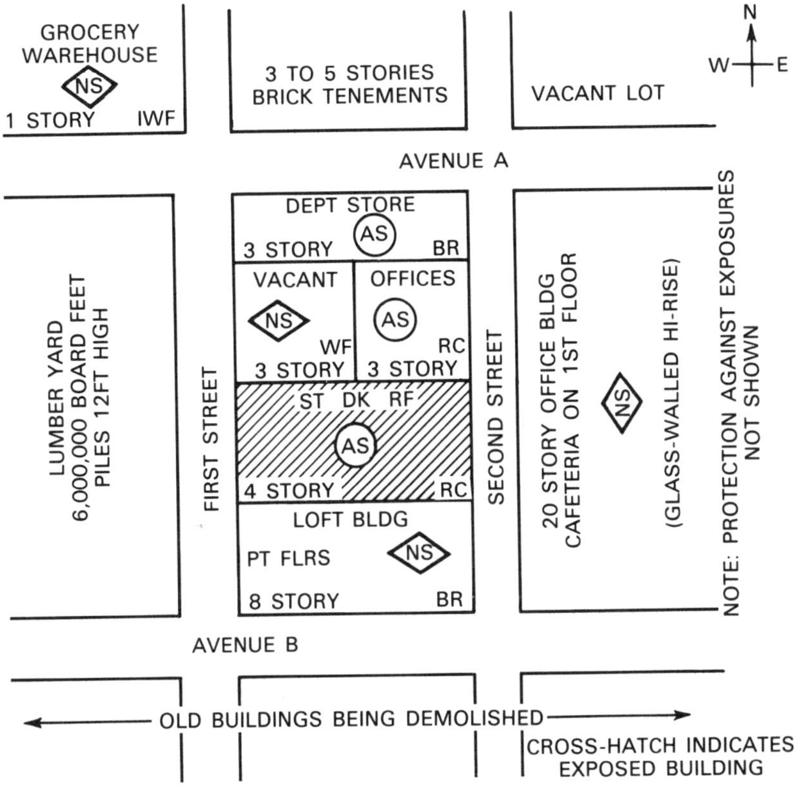

```
                                                              N
                                                              ↑
  GROCERY                                                 W ──┼── E
  WAREHOUSE        3 TO 5 STORIES
     <NS>          BRICK TENEMENTS       VACANT LOT
  1 STORY  IWF
                        AVENUE A
                    ┌──────────────┐
                    │  DEPT STORE  │
                    │  3 STORY <AS> BR │
                    │ VACANT │OFFICES│       20 STORY OFFICE BLDG
  LUMBER YARD       │  <NS>  │ <AS> │       CAFETERIA ON 1ST FLOOR
  6,000,000 BOARD   │   WF   │  RC  │       (GLASS-WALLED HI-RISE)
  FEET PILES        │ 3 STORY│3 STORY│            <NS>
  12FT HIGH         │////ST DK RF///│
                    │////  <AS>  ///│       NOTE: PROTECTION AGAINST
                    │ 4 STORY    RC │             EXPOSURES NOT SHOWN
                    │  LOFT BLDG    │
                    │  PT FLRS <NS> │
                    │  8 STORY   BR │
                    └──────────────┘
                        AVENUE B

  ←──── OLD BUILDINGS BEING DEMOLISHED ────→
                                    CROSS-HATCH INDICATES
                                    EXPOSED BUILDING
```

such as that shown in Figure 3. To the north of the central building, which is marked by diagonal lines, stand an adjoining, unsprinklered, vacant, wood frame building and an adjoining, sprinklered, reinforced-concrete office building. Both adjoining buildings are three stories high, and the exposed building has four stories, as show in Figure 4. From these data, we can draw the following conclusions.

1. There should be a blank, fire resistive wall between the exposures and the lower three stories of the exposed building.
2. Wall openings in the fourth story of the exposed building

Fig. 4. Building exposure, occupancy layout (no scale)

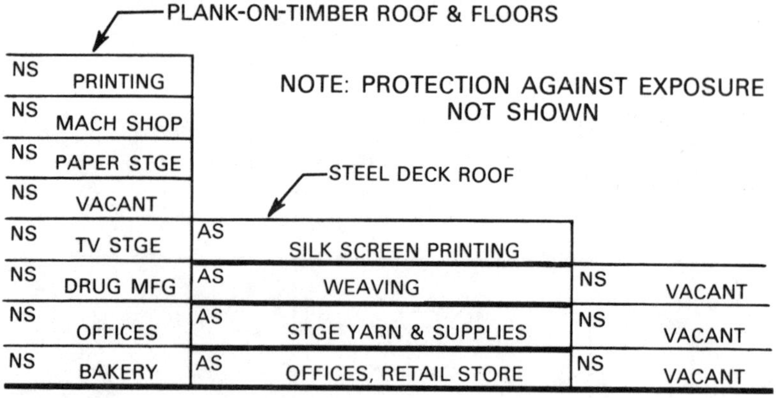

NS PRINTING	\multicolumn{3}{c}{PLANK-ON-TIMBER ROOF & FLOORS}		
NS MACH SHOP	\multicolumn{3}{c}{NOTE: PROTECTION AGAINST EXPOSURE NOT SHOWN}		
NS PAPER STGE	\multicolumn{3}{c}{—STEEL DECK ROOF}		
NS VACANT					
NS TV STGE	AS	SILK SCREEN PRINTING			
NS DRUG MFG	AS	WEAVING	NS	VACANT	
NS OFFICES	AS	STGE YARN & SUPPLIES	NS	VACANT	
NS BAKERY	AS	OFFICES, RETAIL STORE	NS	VACANT	

need no special protection directly above the roof of the office building, but they must be protected over the roof of the wood frame building.

3. To the east rises a twenty-story office building. Although this building might have a major fire, it stands so far from the exposed buildings that wall openings in the exposed building probably need no special protection.

4. To the south of the exposed building stands an eight-story, unsprinklered loft building that has brick walls and plank-on-timber roof and floors. The requirement for some type of blank, fire resistive walls for the first four stories is, or should be, obvious. There is, however, another type of potential exposure hazard that is often overlooked: the possibility that the walls of the top four stories might collapse. Hundreds of tons of brick falling on the roof of the four-story building might do little damage if the roof were of reinforced concrete, but the same weight falling on a steel deck roof could easily destroy the roof and create a severe hazard to life and property in the fourth story. Whether or not such a collapse is likely in the event of fire depends on both construction and occupancy conditions in the loft building.

5. On the west side, the plan shows the beginnings of a large lumberyard with fairly high stacks of lumber. To properly ap-

praise the hazard potential from the lumberyard, an inspector must know the width of the separating street, the distance between the street and the lumber, pile spacing, and lumber sizes. All these factors help to determine what degree of protection the exposed building needs.

C. Severity of exposure. After determining what the exposures are, an inspector must calculate their severity. To do this requires consideration of many factors, including certain features of the exposing and exposed buildings. Principal factors, not necessarily in order of importance, are as follows:

1. Exposure factors

   a. Temperature developed by exposing fire. (Various materials burn at different temperatures.)

   b. Floor loads of combustibles in exposing building and yard loads of combustibles in nonbuilding areas.

   c. Total mass of combustibles likely to be involved in a single exposing fire and the potential heat production therefrom.

   d. Type of construction of walls, roofs, and floors of exposing building.

   e. Potential duration of exposing fire.

   f. Wind direction and velocity at time of fire. (Potential can be estimated in advance.)

   g. Humidity.

   h. Accessibility of exposing fire to fire fighting forces.

   i. Extent of installed fire protection in exposing buildings.

   j. Distance between exposing and exposed buildings.

   k. Height of exposing building relative to height of exposed building.

   l. Presence or absence of intervening noncombustible construction.

2. Exposed building factors

   a. Type of construction of walls, roofs, and floors. (Particular attention should be paid to walls facing or adjoining those of an exposing building.)

   b. Area of openings in exposed walls relative to total wall area. (Consider a combustible wall as being 100 percent openings.)

   c. Extent of installed protection systems.

   d. Protection of wall openings.

   e. Occupancy of exposed buildings. (Occupancy materials differ in autoignition temperature and in sensitivity to damage by heat, smoke, or water.)

D. Burning intensity rate. Once an inspector has obtained information on those of the above items that are appropriate, he can take one of the final steps to determine the actual severity of exposure: establishment of a burning intensity rate class.

1. Classifications. There are a number of systems for classifying burning intensity rates. The most common includes four categories: slight, moderate, considerable, or severe. Some examples of each category are as follows:

   a. Slight. Noncombustible or fire resistive schools, hospitals, and office buildings; light brush.

   b. Moderate. Apartment buildings having noncombustible walls spaced at least 15 ft. apart, with fire partitions at appropriate intervals, and no more than twenty dwelling units; schools, hospitals, or other buildings that have noncombustible walls and no more than four stories or no more than 10,000 sq. ft. of floor space; diked, flammable-liquid tanks having a combined capacity of not over 30,000 gal.; lumberyard totalling 250,000 bd. ft. or less.

   c. Considerable. Warehouses having four stories or less, noncombustible walls, and a floor area of 10,000 sq. ft. or less; wood frame institutions with single floor areas of not over 5,000 sq. ft., small manufacturing buildings with mostly ordinary hazard occupancy and noncombustible walls.

   d. Severe. Multistory, multiunit wood tenements; warehouses and manufacturing plants with high hazard occupancies and areas of more than 10,000 sq. ft.

2. Variations. The burning intensity classes cited above are guidelines only and should be used with judgment, giving due consideration to the inevitable variations in buildings and occupancies that will be found in the field. For example:

   a. Noncombustible walls may range from iron on steel frame to masonry, and they may or may not have windows and doors.

   b. Roofs and floors may consist of boards on joists, planks on timber, steel deck, or concrete.

   c. High hazard occupancies may or may not have built-in safeguards. For example, it is not uncommon for tanks of flammable liquid to have safety relief valves that will operate during a fire and dump the liquids through a trapped drain into a buried or otherwise protected recovery tank.

   d. In general, buildings protected by automatic sprinklers that are in service and have an adequate water supply do not create

Fire Exposure 55

a serious hazard to other property. The keys here are adequacy and reliability.

(Note: Occupancy classes as given in NFPA Standard 13 provide good guidelines for hazard classification.)

E. Degree of exposure. Once the burning intensity class has been determined, it becomes possible to evaluate the degree of exposure. Figure 5 shows one type of evaluation chart for determining the degree of exposure. This chart includes headings for the four major burning intensity rates: severe, considerable, moderate, and slight. Clear distances between exposed and exposing property are indicated on the vertical scale, and the degree of exposure is indicated on the horizontal scale. As an example of how to use this chart, assume an exposing building with a severe burning intensity rate located 60 ft. east of the exposed building. Assume also a similar building 180 ft. west of the exposed building. Degrees of exposure are determined as follows:

1. At the 60-ft. mark on the vertical scale, draw a horizontal line to its intersection with the severe-burning-intensity line, and drop down vertically to the degree-of-exposure line. The intersection is at 4.25, indicating a severe degree of exposure.

2. Use the same method at the 180-ft. mark. The horizontal line intersects with the degree-of-exposure line at 17.5, indicating the exposure is slight.

Figure 5 or a similar chart should only be used to determine degree of exposure in average cases. It is not intended for evaluating unusually large, high-burning-intensity occupancies such as the lumberyard shown in Figure 3.

F. Duration. Once the degree of exposure is known, only one step remains before the inspector can decide what exposure protection is needed. This step is to estimate the potential duration of the exposure fire in question. There is little point in providing four hours' protection against a forty-five minute fire potential, nor in providing an hour's protection against a possible four-hour fire. The goal is to provide what is needed plus a reasonable margin for safety.

G. General types of exposure fires. External exposure fires fall into three basic classifications: local, conflagration, and fire storm. These may be defined about as follows:

1. Local exposure is that created by a fire in the immediate vicinity of the building under consideration. The exposing fire may be in an adjoining property, in a building across the street, or in other nearby property. In general, local external exposure fires

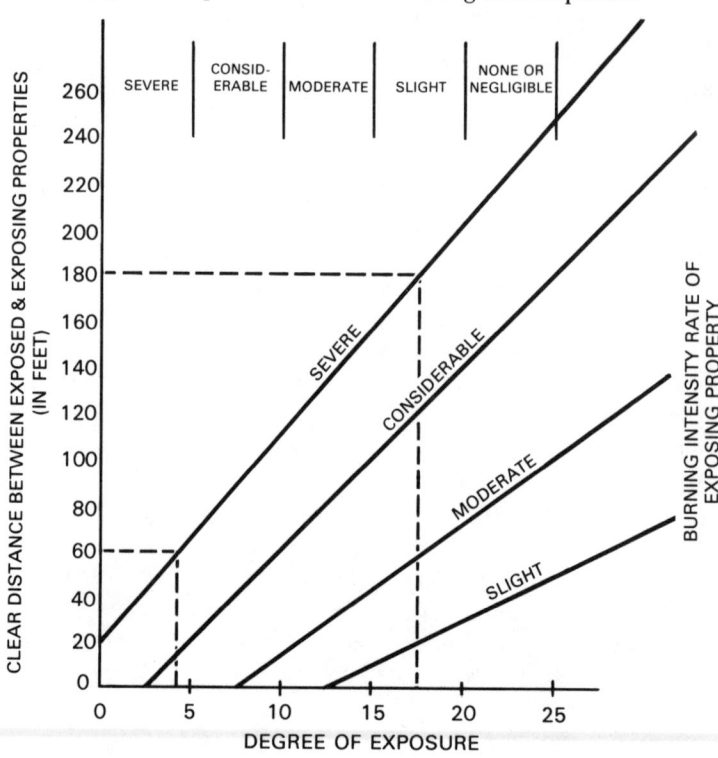

Fig. 5. Sample chart for determining local exposure

are those that originate in property within 250 to 300 ft. of the exposed building.

2. Conflagration exposure has been defined in many different ways. For the purpose of this study, it is perhaps best to define such an exposure in broad general terms, as follows: A conflagration exposure is that created by a fire in buildings or other property so arranged that a fire originating there is likely to involve many buildings and cover a wide area. Excluded from the classification are:

    a. Single industrial plants with many buildings, unless a fire in the plant is likely to involve other properties.

    b. Closely spaced groups of storage or mercantile buildings, unless fire is likely to jump from such a group to other properties located across natural or prepared fire breaks.

3. Fire storm is a special category and can only occur where certain definite conditions are present. In one sense it is a conflagra-

Fire Exposure 57

tion, but an added hazard is present. In a fire storm, air moves towards the center of a group of fires, and the air velocities reach hurricane speeds. At the center, the combined winds form a column of air and toxic gases that rises at a tornadolike velocity. Fire storms generally result from war attack and are outside the scope of normal fire prevention activities. For those students who may be particularly interested, a reading of an unexpurgated edition of *Fire and Air War* is recommended.

II. Defenses against exposure.

A. Defenses against conflagrations. Most fire prevention personnel are primarily concerned with developing specific, economically feasible means for preventing the spread of fire from one property to another. Only a few individuals with special responsibilities are chiefly concerned with prevention of conflagrations as such. All, however, should be aware of the basic means of preventing conflagrations and lend such support as may be appropriate to proposals for defenses against conflagrations. Some of the principal defenses are as follows:

1. Systematic city planning that provides for high standards of construction and protection, zoning to control location and size of high-hazard occupancies, wide streets and other natural fire barriers, construction of physical fire barriers at prescribed intervals in congested areas, and other methods of fire prevention.

2. Widespread installation of fire protection systems with adequate water supplies and provision of manual fire-fighting capabilities.

3. Provisions for the removal or control of unusually hazardous operations.

4. Adoption of building codes and ordinances to enforce the measures described above.

5. Protection of individual buildings. Devices that protect individual buildings from potential conflagration exposures include those primarily installed for protection against local exposure. For example, a sprinklered, reinforced-concrete cotton warehouse with fire shutters at the windows satisfactorily withstood a conflagration that destroyed 1,600 buildings, including those on all sides of the warehouse. Damage inside the warehouse was minor.

B. Defenses against specific exposures. In general, defenses against exposure fires in adjacent or adjoining property should include one or more of such protective devices as blank walls; wired glass windows with or without sprinklers; sprinklers over plain windows; fire shutters; and fire doors. The selection of devices depends on the potential exposure.

1. Blank masonry walls are needed where exposures are severe. Fire resistive ratings selected should be based on the anticipated duration of the exposure fire. (A two-hour wall is useless, or nearly so, against a five-hour fire. In contrast, a five-hour wall to protect against a one-hour exposure would be unnecessarily expensive.) When blank walls are used, it is necessary to bear two considerations in mind.

    a. When exposing and exposed adjoining buildings are of approximately the same height and roofs are combustible, masonry common walls must extend at least 3 ft. above the roofline to prevent fire from jumping the barrier.

    b. When exposing and exposed buildings have wood walls, one of three methods should be used to prevent fire from bypassing the fire wall.

        1. The separating masonry wall should extend at least 3 ft. beyond the wood walls.

        2. The masonry wall should be winged for 5 ft. on each side of its junction with the wood walls.

        3. One of the wood walls should be replaced with masonry for 10 ft. at each end of the masonry wall.

2. Wired glass windows with outside open or automatic sprinkler systems are used where the exposure hazard is considerable or, in some instances, severe but the anticipated duration of serious exposure does not exceed the resistance capabilities of the defenses provided. In setting up such defense arrangements, it is necessary to recognize the following facts.

    a. Not all wired glass is alike in its resistance to fire. Selection should be based on UL or FM ratings of effective resistance time.

    b. It is possible to provide some security by replacing plain glass in wood sash with wired glass in wood sash. Such arrangements should be used with judgment.

    c. Open sprinklers to be effective require that someone be present to turn them on. Plants that operate part-time and lack watchman service should use automatic sprinklers.

    d. Automatic sprinkler systems must have some means to collect heat to assure timely operation. In cold weather areas, such systems should be nonfreeze or dry pipe.

3. Wired glass or glass block panels without sprinklers are used

Fire Exposure 59

where exposures are less severe. Either is acceptable, although glass block may shatter under severe fire exposure.

4. Open or automatic sprinklers systems may be used as appropriate over plain glass windows. Whether used in this manner or with wired glass, water curtains alone will not stop the passage of radiated heat. Accordingly, where an exposure fire would produce considerable radiated heat, a physical barrier is needed in addition to or in place of the water curtain.

5. New installations of fire shutters are seldom recommended today. However, many thousands of fire shutter installations are still in service, and these need to be considered in evaluating a building's protection. Properly maintained tin-clad wood fire shutters can provide effective protection. All-metal shutters, however, are ineffective because they provide little insulation.

6. Fire doors of appropriate fire resistance are available for the protection of door openings in exterior walls.

C. Selecting defenses. To determine what fire defenses are needed for any particular situation, it is first necessary to determine the degree of exposure and to estimate the anticipated duration of exposure. On the basis of this information, an inspector can determine which of the various types of fire defenses should be utilized. Figure 6 provides illustrations of some of the major factors involved in selecting defenses.

1. When an exposure is of sufficient severity to warrant exposure protection, it is not enough to protect only the wall directly opposite the exposure. Protection must also be provided beyond the end of the exposing building as indicated in Part 1 of Figure 6. Extent of additional protection required varies inversely with the distance between the buildings.

2. Exposed and exposing buildings do not necessarily have parallel walls. There may be setbacks, or walls may be at an angle to each other. The resulting variety of distances between buildings may lead to a need for variations in fire defenses. Part 2 of Figure 6 illustrates some of the variables that may be involved. It is essential that fire prevention personnel learn, through experience and reading, how to judge exposures and what can be done to eliminate or minimize their effect.

3. In addition to the exposure situations shown in Figure 6, consideration should be given to the possible need for protection of wall openings in stories above the exposure. If an exposing building has a combustible occupancy, a fire may break through the

Fig. 6. Exposure protection problem areas

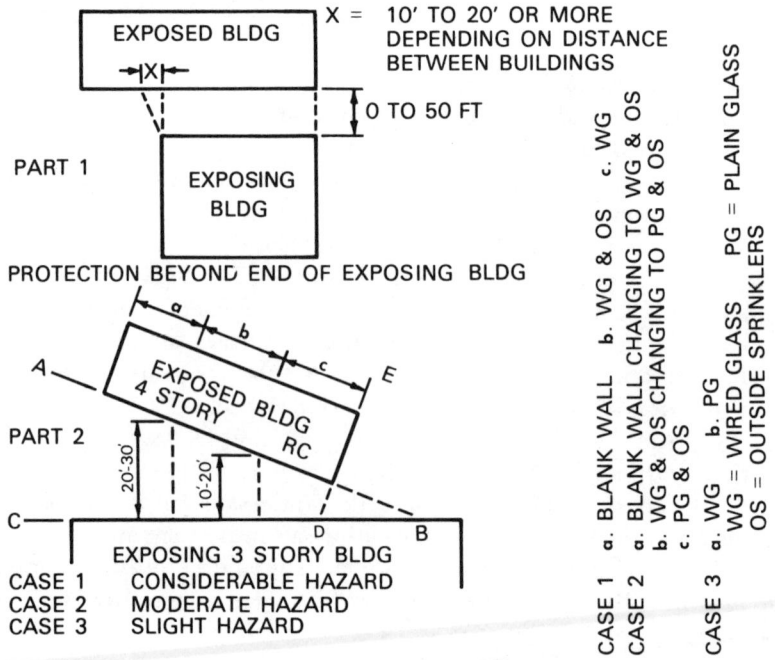

roof, whether it is combustible or noncombustible. The extent to which wall openings above the roof in question need protection, therefore, is directly related to the potential severity of an exposing fire.

Internal Exposures

Internal exposure is the exposure created by occupancies in stories above and below and in areas adjoining the story area under consideration. While the term *internal exposure* as commonly used refers to exposures created by those other than the occupants of the area being considered, the same general line of reasoning should be used for safeguarding individual occupancies having a common owner.

I. Judging the danger. There are many ways of judging potential internal fire exposures. For present purposes, however, it is probable that the logical first step is to consider the susceptibility of the occupancy being studied to fire, smoke, water, and toxic gases. Next, an

Fire Exposure 61

inspector should check on the likelihood that adjacent fires might create such undesirable conditions and determine the probable frequency and duration of such occurrences. Finally, he should evaluate the adequacy of existing or proposed separations between occupancies and determine what, if any, additional fire defenses are needed.

The position of the exposed property relative to the potential exposure fire must be recognized as a factor of major importance. Consider, for example, an occupancy susceptible to water damage. Stories above and below have occupancies likely to have frequent small fires with consequent use of water for extinguishment. The occupancy below poses no problem, but the occupancy above can create a serious problem if the intervening floor leaks readily. As another example, consider a fire that produces unusually toxic gases. In a building with tightly sealed off areas or with zoned air conditioning systems that may be operated on 100 percent fresh air, the gases might pose little danger to building occupants. But in a building through which gases can readily spread, a severe hazard to life might result.

Figure 4, designed primarily for use with external exposure problems, provides a theoretical arrangement of occupancies for one type of textile operation. If the silk screen printing operation shown on the fourth story has few combustibles present, the likelihood of a fire is remote, and, with sprinkler protection in service, it is unlikely that much water would be used for extinguishment. If the fourth story has a concrete floor, there is no reason to anticipate serious water damage to the weaving operations on the third story.

If, on the other hand, the fourth story were used for lacquer manufacturing or for grinding of combustibles, there would undoubtedly be more frequent fires. In addition, substantial quantities of water would probably be used for fire extinguishment; there could be an explosion that would shatter sprinkler piping. Under these conditions, an intervening concrete floor by itself is not enough. Unless the floor can be made reasonably watertight, there must also be such protective devices as waterproof canopies over sensitive items.

There are so many combinations of occupancies that no one book can analyze all the possible exposure problems. Expertise in this field comes only with time—from experience, from regular reading of instructive fire reports, and from other forms of education. But the beginning fire prevention practitioner can learn to evaluate occupancies as to their susceptibility to damage from fire, heat, smoke, water, or toxic gases. He can also learn to judge the likelihood that fires in adjoining occupancies will produce damaging amounts of smoke, water, or toxic gases, and to estimate the frequency with which such fires might occur.

II. Safeguards against fire, heat, smoke, water and toxic gases.
  A. Kinds of safeguards.
    1. Barriers. The usual barriers to the spread of fire consist of such physical blocking devices as fire walls, fire partitions, stair and elevator enclosures, and fire doors at communicating openings. Some equally satisfactory alternatives are outlined below.
      a. In lieu of a fire wall, a building may have two lines of open sprinklers controlled by a deluge valve and heat-actuated devices. Such an arrangement is used in many large assembly buildings where the presence of walls would interfere with operations. The system must have an adequate water supply that is separate from the building sprinkler supply.
      b. Conveyor openings in fire walls may be protected by special spray or sprinkler systems.
      c. There are various special protection arrangements for escalators, as detailed in Chapter 3.
    2. Safeguards for communicating openings. The larger fire barriers usually receive attention, but too little notice may be given to horizontal or vertical communicating openings. The most common of these are as follows:
      a. Pipe and wiring chases. Such openings probably had fire stops as original equipment. Too seldom is a check made to see if the stops are still there after maintenance or alterations work has been done.
      b. Ducts. Most communicating openings in firewalls or in floors have fire dampers, but the fire prevention inspector should check to see whether or not the dampers have been maintained in operating condition.
      c. Pipe sleeves. Some piping through floors and walls is not in chases but passes through sleeved openings. Spaces between pipes and sleeves should be packed with noncombustible material.
    3. Venting devices. Safeguards against smoke and toxic gases normally consist of venting devices. These devices include, but are not limited to, those listed below.
      a. Vents for stair and elevator shafts.
      b. Smoke hatches in roofs of certain types of warehouses and industrial plants. These hatches should preferably be automatic in operation.

Fire Exposure 63

      c. Air-handling systems that pick up smoke and toxic gases and discharge them outside the building upon the operation of automatic or manual controls.
    4. Other devices.
      a. Venting capabilities alone are not sufficient for some types of occupancies. Hospitals and similar organizations need smoketight fire cutoffs.
      b. Protection against water damage is primarily accomplished by the provision of watertight floors, wall scuppers, trapped floor drains leading to safe locations, or waterproof canopies over water-sensitive operations in stories below.

B. Selecting safeguards. Because of the multiplicity of types of occupancy and types of construction, determination of essential safeguards must be made on an individual basis subject to specific code requirements. However, the inspector making such determinations should always remember the following considerations.
    1. Recommendations should be for the situation at hand and should be cost-effective. "Gold plating" is self-defeating.
    2. Quite often it is possible to select a safeguard that meets more than one need. For example, a smoketight fire door is simpler to maintain and cheaper than an ordinary fire door plus a smoketight door, and the two types of protection are equally effective.

Standard automatic sprinkler systems and certain other installed protection systems also provide protection against internal exposure. However, these systems do a great deal more than protect against exposure and will accordingly be covered separately (see Chapter 6).

chapter 5
Building Utility Systems

Definitions

Most building codes define building utility systems as those systems that are essential to the functioning of a building for its designed purpose and that normally remain with the building if the owner or tenant moves out with or without his machinery or other equipment. Building utility systems include all parts of the utilities up to and including the tie-ins to the corresponding utility systems that have been or will be installed to serve particular equipment or operational needs. Building utility systems do not, in general, include any part of the auxiliary systems.

Some building codes cover auxiliary utility systems to some extent, but these requirements for the most part are used in conjunction with the more detailed requirements set forth in the appropriate fire codes. Under normal circumstances in most areas, auxiliary occupancy utility systems do not come within the direct purview of the building department, although they may do so. Auxiliary utility systems should be classed as parts of building utility systems when the building is so designed that it can be used for no other purpose than the original type of occupancy without major building revisions.

Some examples of the distinction between building and auxiliary utility systems are listed below.

1. Elevators. Passenger and freight elevators, including the power supplies thereto, are almost always 100 percent a building utility. Exceptions to this rule are few.

II. Gas. Supply mains and distribution piping are building utilities. Piping, valves, and other equipment directly serving a piece of equipment or a particular operation are auxiliary systems from the point where they tie in to the basic building system. Systems supplying heat and power to the building are exceptions to this rule, because these units are integral parts of the building.

III. Air conditioning. A system installed for the building as a whole is a building utility. Separate systems, such as those used to supply computer systems or dust-free rooms, are auxiliary systems. Air conditioning systems designed for fire control at escalators form an exception to this rule. Since the escalators normally remain with the building, services to them are also part of the building.

IV. Other types of utility services are divided along the same lines. All building utility systems and auxiliary systems have one fire safety requirement in common. Whenever the utility design requires piercing a necessary fire barrier (walls, floors, roofs), the design should be such that the essential integrity of the fire barrier is maintained. An inspector should be especially careful to see that poke-through assemblies do not breach a fire barrier.

One of the main reasons for separating building systems from auxiliary systems is that a change in a building system is of a more or less permanent nature and usually affects more than one operation. On the other hand, installed equipment is subject to frequent modification as new processes and methods are developed, and changes normally affect only one piece of equipment or several pieces of the same type. As a result, a fire prevention consultant must take substantially different approaches to fire problems in the two types of systems. In each case, however, he or she should be looking for cost-effective solutions.

Electrical Systems

Of primary importance in today's style of living is the continuity of electrical power. Without such continuity, the problems and costs of running the present industrial establishment—or, for that matter, any important part of the urban social system—would be almost impossible to handle. Doubters of this fact are referred to the effects of recent major blackouts in large urban areas. These blackouts are but multiplications of what happens to an individual plant or institution when power is lost for an appreciable period of time.

Maintaining electrical systems over a wide geographical area is a field for specialists and, accordingly, will not be covered in this text. The fire prevention inspector deals with individual plants, institu-

Building Utility Systems 67

tions, and other buildings. In addition, his concern is not with the electrical design of the system but with its safety. He must make certain that the proper types of protection equipment have been installed and are being maintained and that adequate fire protection is present.

I. Transformers. There are three types of transformers in general use: oil-insulated, askarel-insulated, and dry. All types should be provided with such protective devices as fuses and circuit breakers on the primary side or with a satisfactory equivalent thereof. All transformers should also be protected by lightning arrestors if there is a likelihood of lightning surges. The inspector should satisfy himself or herself that protective devices were prescribed by qualified persons and are being maintained.

There are standard checklists for essential tests and maintenance for each basic type of transformer. Inspectors should make certain that these tests and maintenance procedures are being carried out and that corrective actions are promptly initiated when deficiencies occur.

Most transformers currently used in the United States are three-phase, and replacements or substitutes for them are available fairly readily. However, there are some instances of two-phase equipment. It is important to determine the availability of substitutes or replacements for such equipment in the event of a transformer fire. If a disabled transformer supplying a key operation is not quickly replaced, the cost of prolonged crippling of plant operations could far outweigh the fire loss to the transformer.

A. Oil-insulated transformers. Fires in this type of tranformer result primarily from insulation breakdown, low oil level or oil deterioration, failure of insulating bushing, lightning, or overloading. These and other deficiencies should not occur if plant testing and maintenance are satisfactory. In general, plant procedures are good, as evidenced by the excellent overall transformer fire loss record. Nevertheless, there have been some disastrous transformer fires that could have been minimized by providing the proper safeguards against the spread of fire. Precautions that can be taken as conditions warrant include, but are not limited to, the following:

1. Provide a clear space of 25 ft. or more between transformers and building walls, unless walls are blank masonry.
2. Provide similar clear spaces between important transformers, unless noncombustible barriers with suitable fire resistance are provided between units.
3. Provide fixed and/or portable water spray protection systems as size and importance of units warrant.

4. Confine burning oil that might overflow. There are many methods of doing this. One of the most common is to mount the transformer above crushed stone on a curbed concrete pad with a trapped drain leading to a safe location.

5. Locate fire resistive barriers between transformers and walls where proper spacing cannot be arranged and where for operational or other reasons it is not feasible to have a building wall of blank masonry or concrete.

6. If a transformer must be located on a combustible roof, mount the device above a curbed concrete mat with suitable oil drainage facilities.

7. Provide suitable fire resistive vaults for transformers whose size and potential hazard to other property warrants doing so.

B. Askarel-insulated transformers. The insulating liquid in askarel transformers is nonflammable. Internal arcing, however, may generate gases. Installations should provide means for absorbing these gases or for venting them outdoors.

C. Dry transformers. Insulation failures at windings are responsible for most fires in dry transformers. Such fires are usually confined to the unit of origin and are readily controlled with portable fire equipment. For larger units—110 kva or more—kept indoors, it is desirable to have a noncombustible transformer room unless the building is sprinklered or the occupancy near the transformers is essentially noncombustible.

For various reasons that are not the concern of fire prevention personnel, nearly all outdoor transformers and many of the larger indoor units are oil-insulated types. At each location, an inspector must use judgment in determining the threat to operations of a potential fire and deciding how much protection is needed to eliminate the hazard or to bring it within acceptable limits. For example, a plant might have five 1,000 kva units and an actual workload of 3,500 kva. Loss of one unit would not seriously interfere with plant production. Another plant with the same transformer capacity might have a workload of 4,900 kva. Loss of a unit at that plant could conceivably cripple operations for an extended period. The difference in threat to operations would help decide whether or not special fixed protection should be provided.

At indoor installations, a reverse consideration is possible. A combustible building or a building with combustible occupancy and without sprinklers could jeopardize transformers serving other buildings. In such a situation, the transformers should be in suitable vaults.

II. Generators and motors. Generators may be divided into two

Building Utility Systems 69

classes: (1) those generating primary power and (2) those supplying auxiliary or emergency systems. Protection of primary power generators will be covered below, under the heading Powerhouse and Switchgear. Of primary importance with auxiliary system generators is their ability to come on line instantly in the event of primary power failure. Many auxiliary generators supply such vital systems as emergency lighting, police or fire alarms, and intensive care facilities.

Electric motors are, of course, found in all types of industry, in many mercantile or institutional establishments, and in many residential occupancies. There are hundreds, if not thousands, of types and sizes of motors, many of them designed for special hazards or occupancies. Such motors are not interchangeable with other motors of the same size but a different hazard rating. For fire prevention purposes, it is not necessary to have detailed electrical design knowledge. It is, however, essential that fire prevention personnel know the few basic types and their proper areas of usage.

A. Generators for auxiliary systems. These generators are usually driven by internal combustion engines, either gasoline or diesel. Some natural gas engines are also in use. Primary areas of concern include those listed below.

1. Presence or absence of satisfactory means for automatic switchover from primary power when that source fails. Too often it is forgotten that primary electric power is often lost during a fire. Failure of an auxiliary system to take over emergency lighting or critical operations can turn a minor fire into a disaster. (Fire prevention includes limiting the spread of fire.)

2. Proper location and venting of storage systems for fuel to drive generator engines (see Chapter 9, Flammable Liquids and Gases).

3. Proper installation and maintenance of fuel systems, including such ventilation as might be required for the type of fuel used.

4. Protected location of engine-driven generators, whether indoors or outdoors. For some situations a cut off, noncombustible or fire resistive room or shed would be in order.

B. Motors. Electric motors in a wide variety of sizes and types are used throughout almost all varieties of activities. Detailed studies of electric motors are not essential for the fire prevention generalist. It is, however, vital that fire prevention personnel be able to determine whether or not a motor is suitable for the location in which it is installed and whether or not it is being properly maintained. Precautions that should be observed include, but are not limited to, the following.

1. Hazardous locations. The three basic types of hazards are Class 1, flammable vapor or gas; Class 2, combustible or conductive dust; and Class 3, easily ignitable lint. Name plates on motors indicate the hazard for which the device was designed. It is important to realize that a motor designed for one class is not suitable for the other two. Also, each class is divided into two divisions. Division 1 motors are designed for use where the hazard is normally present, and Division 2 motors are for use where the hazard is not normally present. Accordingly, Division 1 motors may be used in Division 2 locations of the same class, but Division 2 units are not acceptable in Division 1 locations.

2. Lubrication. Unless lubrication is done properly and on a regular schedule, there will inevitably be overlubrication or underlubrication, either of which can result in fire.

3. Cleaning. There should be a cleaning schedule for all open motors where lint or other light, combustible debris is present. Some laundries and garment factories are prime examples of lint-producing operations.

4. Overcurrent protection. All motors of high value or of vital importance to operations should be equipped with overcurrent protection.

5. Vibration. Where flexible connections are necessary and excessive vibration is present, as in weave sheds of textile plants, only connections approved for such use should be provided.

6. Insulation. Many motors are installed in oily locations or where acid fumes are present. Insulation in such locations should be resistant to oil or acid as appropriate.

The student should note that many of the comments on generators and motors in building utility systems are also applicable to the same types of equipment in specific occupancy situations.

III. Powerhouses and switchgear. The vast majority of today's industrial plants, mercantile establishments, and institutions receive their electrical power supplies from utility networks, but a number of the larger industrial plants generate their own power. Virtually all large facilities have switchgear rooms or buildings. Fires or explosions in powerhouses or switchgear can not only result in extensive physical damage; they can and have crippled or stopped manufacturing and commercial activities for extended periods of time. It is entirely possible for a $50,000 fire in a powerhouse or in switchgear to result in millions of dollars worth of loss in production.

Building Utility Systems

A. Powerhouses. The fire and explosion potentials at powerhouses depend primarily on building construction and the method used to drive generating equipment.

1. Construction. New powerhouses and additions to old units should be of fire resistive or noncombustible construction, and protection against exposure should be provided as necessary. Existing combustible powerhouses should be improved by such methods as flameproofing the underside of the roof deck, replacing combustible roofing, and installing noncombustible wall insulation.

2. Combustibles. In powerhouses, such materials should be kept to an absolute minimum, and appropriate protection should be provided where needed. For example, a powerhouse may require a separate room for combustibles and possibly room sprinkler protection.

3. Fuel hazards. Such hazards in powerhouses are similar to those found in boiler plants and will be covered in the Heating Systems section of this chapter.

B. Switchgear. For the purposes of this text, switchgear refers to main or important power and light switchboards and equipment directly associated with such switchboards. Important fire prevention considerations for switchgear are listed below.

1. Location.

   a. Switchgear equipment suitable for hazardous locations is not available. Switchgear should not, therefore, be located where flammable vapors or combustible dusts are likely to be present.

   b. Locate switchgear so that personnel would have adequate means of egress in event of an emergency.

   c. Locate indoor switchgear in ventilated, fire resistive, unsprinklered rooms. In multistory buildings, be sure the floor above is waterproof or the equipment is otherwise shielded from potential floor leakage.

2. Combustibles. Such material should be severely limited in switchgear rooms.

3. Insulation. Materials used to insulate bus bars or bus bar joints should be noncombustible.

4. Circuit breakers. These devices should be installed as appropriate. Because of the oil fire hazard, air type breakers are preferable at locations near the panels.

5. Grouped cables. Such cables should be flameproofed or in-

stalled in separate conduits. For fire purposes, lead-covered cables should not be classed as noncombustible. Lead sheathing melts at a relatively low temperature, and the insulation below will burn.

IV. Control devices. The principal types of control devices seen by a fire prevention inspector are switches, circuit breakers, controllers, and fuses. For all these devices, several considerations, listed below, are especially important in fire prevention.

A. Location. If control devices must be installed in hazardous locations, it is essential that only types designed for the particular hazard involved be used.

B. Maintenance. There are three main kinds of maintenance, which apply not only to control devices, but also to other electrical equipment and, in fact, to all equipment involved in fire prevention.

1. Preventive maintenance consists of regularly scheduled inspections, testing, and overhaul to check out equipment for items that might cause trouble.

2. Irregular maintenance is that done to correct defects before equipment failure occurs. Such maintenance is based on irregular visual inspections and spot checks.

3. Breakdown maintenance is repair work done after failure occurs. It goes without saying that preventive maintenance costs the least and provides for troublefree operations.

C. Defects. Some of the most common defects that have resulted in fires are as follows:

1. Overload. Frequent blowing of fuses or repeated operation of circuit breakers provides clear indication of a possible overload. Another overload indication is a hot or warm fuse box.

2. Jumping. Fuses are sometimes jumped to avoid having to replace them frequently. Coins behind plug fuses and copper strips between contacts for cartridge fuses are the usual methods used to short-circuit fuses. Both are dangerous.

3. Overfusing. More than once, oversize fuses have been installed after overloads started blowing fuses. There are definite capacity fuses for each commonly used size of wire.

4. Breaker tieback. Another, fortunately not common, dangerous practice is to tie back frequently operating breakers so that they cannot function.

5. Compensator tiebacks. Sometimes a starting compensator on a motor controller will be found tied back in the running position. This is very poor practice.

6. Control devices. These also should be checked for proper

Building Utility Systems

operation. Smooth, fast operation does not result when connections are loose.

V. Wiring circuits. Electrical defects and failures rank first among fire causes in terms of dollar losses and second in terms of number of fires. In most types of buildings and occupancies, shortcomings in wiring circuits probably account for well over half the loss from electrical fires. Most of these losses can be prevented by a little knowledge of installation rules and by making certain that a good maintenance program is in effect.

A. Building Codes. At the outset, fire prevention personnel should have a working familiarity with a few basic provisions of the building code used in the area. In general, these key provisions can be summarized as follows:

1. Wiring as well as other electrical fixtures should comply with the provisions of the *National Electric Code.*

2. All materials, fittings, and devices forming part of a wiring circuit should be approved for the proposed type of use by a nationally recognized testing laboratory. Before this requirement is lightly dismissed with a "So what, everybody knows that," it must be remembered that hundreds of millions of dollars worth of electrical work is done regularly by foreign concerns using their country's equipment. The testing laboratories of these nations may or may not be the equipment of the UL or FM laboratories. Unless the laboratories are reasonably comparable, backup approval by UL or FM might well be required.

3. At all new installations and at extensions to existing systems, provision should be made for inspection and approval before any wiring is concealed or any power is turned on.

4. Provision should be made for regular reinspections where public safety is involved.

5. Strict control must be exercised over the authority (usually the electric department) to grant permits for essential limited electric supplies to uncompleted buildings.

B. Other considerations. Beyond these basic items the fire prevention inspectors' operations should be governed by the requirements of applicable fire codes and by an awareness of those defects that are responsible for most electrical wiring fire losses. An individual fire inspector cannot be expected to remember all the provisions of the numerous fire codes. He should, however, remember where in the codes to find special guidance when the need arises. Picking up the electrical defects that cause wiring fires is another matter. To do so,

an inspector needs chiefly habits of careful observation and common sense, as indicated by the following list of defects that result in perhaps 90 percent of all wiring failures and related fires.
1. Warm and hot fuse boxes, which indicate overloaded circuits.
2. Repeated tripping of one or more circuit breakers.
3. Oil-soaked wiring. (Pay particular attention to wiring in cable trays under oil-soaked floors or in floor trenches where oils are present.)
4. Ordinary hazard wiring in hazardous locations or special occupancies—for example, acid rooms, flammable refrigerators, and paint rooms.
5. Tandem cords and multiple outlets.
6. Extension cords that are one or more sizes smaller than appliance wiring.
7. Overlong extension cords.
8. Splices and taps outside of approved conduit boxes.
9. Wiring run on the floor or under carpets. (Such practices may abrade wiring insulation.)
10. Drop wiring from ceiling fixtures not approved for that purpose.
11. Portable lights without wire guards.
12. Circuit breakers with contacts taped closed. (This wouldn't be done unless the breakers tripped frequently.)
13. Wiring secured to walls or ceilings by nails or other substandard fasteners.
14. Grouped combustible cables. (Fire in one can destroy all.)
15. Important wiring strung over combustible roofs. (A roof fire could knock out power to fire pumps or other key operations.)
16. Jumpers across cartridge fuses, wrong sized fuses, or metal discs or coins back of plug fuses.
17. Loose fittings and supports, particularly where vibration is present.
18. Excessive temporary wiring.
19. Cracked outlets or receptacles.
20. Brittle or frayed insulation.
21. Corrosion of conduit, tubing, or boxes.
22. Improperly grounded circuits.
23. Cabinets or boxes full of dust or lint. (Where this occurs, doors or covers are usually missing or open.)
24. Loss of tension on cartridge fuse clips.

25. Broken lamp holders or tube holders.
26. Fluorescent light ballast units against combustible ceilings.
27. Lack of guards to prevent mechanical injury.
28. Wiring channels that lie under water lines but lack provision for drainage in the event of leakage.
29. Sharp, concealed curves in armored cables with vibration present.
30. Wiring in the same trench with piping for flammable liquids or gases.

It should be noted that, for all practical purposes, no specialized engineering training is required to detect any of the above defects or, for that matter, most of the other electrical defects previously noted. An inspector does need developed powers of observation and the judgment to recognize the potential for loss should a fire occur. For example, how much damage can a short circuit in a lighting circuit do in a fire resistive building used for the storage of structural steel? The answer, of course, is very little. On the other hand, the same short circuit in the carding room of a cotton mill could result in a major fire. Good judgment in handling a situation goes a long way towards securing cooperation from property owners and others.

Air-Handling Systems

One of the major technical advances of our times is the development of many types of effective and efficient air-handling systems. These developments have, in addition to making life more comfortable, made possible many highly desirable and important industrial processes, new and advanced types of buildings, and better handling of various kinds of mercantile and institutional operations.

The term air-handling systems, as commonly understood, includes all duct systems moving air by mechanical means, except for certain special systems that are specifically designed to move or remove dusts, stock, refuse, or flammable vapors. For the purposes of this text, air-handling systems are considered to include cooling, heating, ventilating, and air conditioning systems or any combination thereof.

I. Hazards. All types of air-handling systems, regardless of size or designed purpose, may spread fire, smoke, toxic gases, and heat throughout the building or area served by the duct system. Examples of such events are listed below.

A. A small fire produces smoke particles which, because of missing and/or poorly maintained safeguards, are transmitted by the duct system to an area containing sensitive electronic equipment, expensive pharmaceutical products, or other occupancy susceptible to smoke damage.

B. A small fire occurs in a chemical laboratory, and highly toxic gases are transmitted by the air-handling system to sleeping quarters.

C. A small fire in a poorly installed and maintained system picks up radioactive particles and transmits them through the duct system. Radioactive contamination of adjoining and adjacent areas results.

Because of actual fire instances such as those noted and of other fire hazard potentials, no fire protection program can afford to ignore the basic requirements for fire safety in air handling systems. The function of fire prevention personnel is not to design the system but to make certain that necessary safeguards are in place, that proper materials were used, and that the system is receiving proper maintenance.

II. Basic controls. All systems should be provided with means for controlling air flow, and fan motors, air intakes, and outlets should have safety devices suitable to the type of system involved. Some systems should also be provided with appropriate fire detection and/or protection devices.

  A. Controlling air flow.

    1. All systems should be provided with emergency manual shut-off controls. In windowless and basement buildings, there should also be an emergency manual control for turning the system on in the event of fire during an air-handling system shutdown. It is customary to shut down an air-handling system during a fire in an ordinary building, but, during a fire in a windowless building, it is usually best to have the system operate so as to vent smoke and heat.

    2. It is desirable that larger systems have controls to change the source of the air supply or to reroute air flows. If a fire occurs where the system normally uses 85 percent recirculated air and 15 percent fresh air, the ability to switch to 100 percent outside air is highly useful. If an exposure fire occurs near a fresh-air intake, the ability to switch to another air supply is needed. If a fire produces large volumes of smoke that cannot be removed satisfactorily by normal means, a system able to reroute air flows becomes extremely valuable. Such a system can pick up the smoke-laden air and spill it outside the building.

    3. Controls such as those just cited should be located in fire safe areas that are readily accessible to responsible employees and responding fire departments but not to the public. In windowless or

basement buildings, it is best to have the controls in a fire rated building or a room with access from the outside only.

B. Protective devices for fan motors. Fan motors should be installed outside of ducts and plenum chambers. Unfortunately, this safety rule is violated at times for one reason or another. A motor in a duct or a plenum chamber should be provided with a protective device that cuts off the current before the motor becomes so hot that it smokes.

C. Air intakes and outlets. These devices are not, strictly speaking, controls. However, the locations, sizes, and number of intakes and outlets play important roles in the overall control of an air-handling system. Some of the most important considerations in the design of these devices are as follows:

1. To the maximum extent possible, both fresh-air intakes and recirculated-air return intakes should be located so as to minimize the likelihood that an outside fire and its products may enter the system. When such hazards are unavoidable, automatic fire doors or dampers are needed at the intakes. Regardless of the immediate exposure, the openings of exterior intakes should be screened to prevent the entry of foreign materials.

2. Many plants and institutional or residential properties make use of liquefied petroleum gas and/or other flammable gases. Where these gases are used, it is essential that intakes be located well away from the gas storage so that leaking flammable gases cannot be drawn into the system.

3. Return air intakes should never be provided in an area where appreciable quantities of flammable vapors are present. Such intakes should also be avoided to the extent possible where combustible dusts are present.

4. Exhaust outlets, which may be used to remove smoke-laden air, should be located away from any fresh-air intake or any other building wall opening so that there is no possibility that the products of combustion will be drawn back into the building.

D. Detection and protection devices. In places of public assembly or in locations where occupancies are highly susceptible to smoke damage, it is good practice to require the installation of strategically placed smoke detectors. These devices may be arranged to shut the system down, operate specific fire dampers, actuate the fire alarm systems, shut down or actuate one or more fans, actuate fire extinguishing systems, or perform any other function that may be desirable from a fire prevention or firefighting standpoint.

In an air-handling system that is properly designed and installed, automatic sprinklers are needed only for the protection of oil-bath filters. Other sprinkler requirements around air-handling systems are generated by building and occupancy conditions and not by the air-handling system itself.

III. Ducts. All duct systems and their accessories should be designed, constructed, and maintained so as to minimize the possibility that fire or fire products may be transmitted from one area to another. Most new systems are designed with due consideration to fire safety, but some are not. In addition, many older systems have serious defects that are not always apparent to a casual observer. Of major concern to fire prevention personnel are the following requirements and defects, which are not necessarily listed in order of importance.

A. All ducts should be constructed of noncombustible materials. Most of them are, but some are not. For example:

1. A metal duct is encased in a fiber glass cover, which in turn has a wrapper of such combustible material as asphalt-impregnated kraft paper. Flame can propagate along that wrapper.

2. A metal duct has fiber glass insulation with no wrapper. The duct is located where oily vapors or combustible dusts are present. Unless a noncombustible wrapper is provided, the vapors or dusts will in time impregnate the fiber glass and make it combustible.

3. A duct of unknown construction is encased in transite (cement-asbestos board). An inspector should make it his business to determine the composition of the duct. The author had a piece of transite removed from a duct in a 300-seat theater. He found that the duct system, which extended more than 400 ft., was constructed of plywood.

B. All duct linings should be noncombustible, and most of them are, except possibly for the adhesives used to cement the lining in place. A few years ago, three workmen were installing fiber glass linings in large ducts for a major office building. The adhesive caught fire, and the workmen barely escaped through a large intake opening where screens had not yet been installed.

C. Flexible connections should be noncombustible.

D. Clearances between ducts and combustible construction or occupancy should always be provided. Clearances are of particular importance where ducts are used for heating purposes.

E. Ducts are frequently installed in enclosures or in concealed spaces above ceilings. If the enclosures or the ceilings below are combustible or there are appreciable combustibles in the concealed spaces, automatic sprinkler protection may be desirable.

F. Virtually all duct systems require cleaning from time to time, frequency being dependent on the occupancy served. (Systems in offices do not require cleaning as frequently as systems serving commercial cooking facilities.) Cleanout openings should be of sufficient size and so located that all parts of the duct system can be cleaned.

G. If ducts pass through combustible walls, partitions, ceilings, or floors, noncombustible insulation should separate the ducts and the combustible construction. If fire-rated construction is penetrated, ducts should be so installed that the fire integrity of the construction is maintained. It may be necessary to install wall sleeves and pack the space between the sleeve and the duct with noncombustible material. In other cases, the duct may be enclosed with suitable fire resistive material.

H. Ducts need automatically closing fire dampers or doors wherever fire can communicate from one area to another through the ducts (see Figure 7). Such fire dampers or doors should also be provided with means for manual operation.

IV. Cooling towers. Many air conditioning systems, particularly those of larger size, require one or more cooling towers to disperse the heat generated by their operation. Such towers may be noncombustible, but for economic reasons a high percentage of them are, and will continue to be, constructed of wood. It is with the wooden towers that fire prevention personnel need to be concerned. If an inspector casually observes a wooden tower in operation, with water cascading down inside it, he may conclude that the tower poses no serious hazard since it is wet. This can be a dangerous attitude to take. Such towers may burn, and their loss may have serious consequences.

A. Combustibility of towers is illustrated by the situations cited below.

1. Except in tropical areas, the systems will probably be shut down for extended periods during cold weather. At such times, the towers will be dry.

2. Shutdowns for strikes, remodelling, or other reasons are generally long enough for towers to dry out.

3. In large cooling towers, particularly those operating on induced draft principles, substantial portions of the interior remain dry even during full-capacity operation. In fact, fire loss records

Fig. 7. Typical fire damper locations

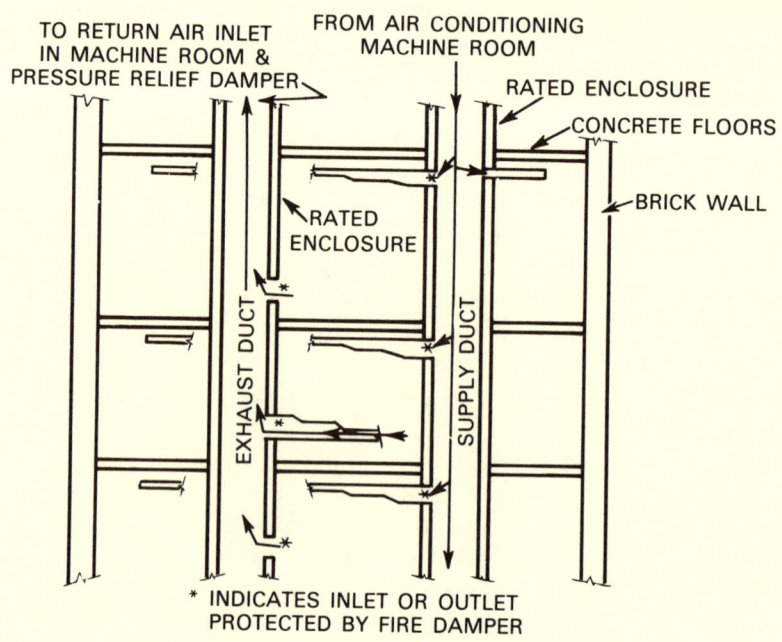

* INDICATES INLET OR OUTLET PROTECTED BY FIRE DAMPER

clearly indicate that from 40 percent to 50 percent of cooling tower fires occur while the tower is operating.

B. The loss of a cooling tower may involve, not only fire damage to the unit itself, but also the interruption of plant operations or even a hazard to life. For example:

1. An air-conditioned hospital is located in desert country. The single, large cooling tower is destroyed by fire. Hospital personnel may then have the problem of caring for critically ill patients in 120° F. heat.

2. A manufacturing operation requires conditioned air throughout most of the plant for satisfactory results. Loss of the cooling tower at such a plant means an operational shutdown even if fire does not enter the building.

C. Safety measures. Most fires in cooling towers result from outside sources, from sparks generated by the fans, or from short circuits in wiring to the fans. To minimize these possibilities, it is desirable to take the following measures.

1. Locate towers well away from such potential outside ignition sources as cupolas, process stacks, incinerators, and other hazardous operations.

2. If a cooling tower fire could result in heavy damage to other property, cause operational shutdowns, or represent a large loss in itself, the tower should be protected as needed by an automatic sprinkler system. In a cold climate, the system would normally be of the deluge type.

3. After a tower has been shut down for any length of time, operating personnel should turn on the water long enough to wet down the inside of the tower before starting the fans.

4. A shutoff switch for fans should be available for use if a fire should occur when the tower is operating.

V. Filters. Many types of air conditioning and ventilating systems are provided with filters to remove dusts, other combustible particles, and/or radioactive materials from the air stream. Filters may be either stationary or moving. Some of the latter use oil baths.

Many types of filters are noncombustible, but it should be recognized that their function is to pick up unwanted particles, which in all probability are combustible. It should also be recognized that cloth-bag filters and oil-bath filters involve combustibles regardless of the nature of the particles they remove. It therefore appears generally desirable to observe the following safety measures.

A. Provide for regular cleaning schedules, and observe a program for periodic replacement of throwaway or renewable filters.

B. Install sprinkler protection where oil-bath and cloth-bag filters are used.

C. Where warranted by the potential fire hazard of surrounding occupancy, provide sprinkler protection.

D. Where radioactive materials are present, prohibit the use of combustible filters.

E. Use noncombustible stationary filter elements to the extent possible.

VI. Zoning. Many modern buildings are so large or so arranged that it is impractical to serve the entire building with a single air-handling system. Such structures are subdivided into zones with each zone having its own air-handling system. A zone system, like a whole-building system, may be designed to operate other building equipment, including fire alarms and fire doors, and has such accessory equipment as is necessary. It is possible to interlock two or more zone systems, but the practice is not common.

Each zone system should of course be properly designed and installed, and information as to what each system does or could do in an emergency should be available. In addition, the location and identification of the controls are important. Preferably, there should be a single master control room. Where a single location is impossible, the number of locations should be kept to a minimum. All control rooms should be fire safe and readily accessible to key employees and responding firefighting personnel but not to the public. In the event of a serious fire, proper control locations can be a major asset not only in limiting the spread of fire, but also in aiding rescue and evacuation of personnel and in controlling smoke damage.

Gas Systems

Prior to 1900, most buildings were illuminated by gas. A substantial number of these properties were also heated by multiple open-flame gas heaters. Today, these systems, for all practical purposes, are no longer in use, having been replaced by more efficient heating and lighting equipment. This change, however, does not mean that the old gas piping has all been removed. Much of it still remains in place because the cost of removal is high. Most of the systems remaining in place have been disconnected outside the building, but some have not. The gas supply to the latter systems was cut off simply by closing the street gas valves, which may or may not be tight.

Modern gas sytems serve many types of uses, including central heating, processing ovens, chemical laboratories, and unit heaters. The pressure in a system may vary from low to high, depending on operating requirements. The variations in pressure may be great enough to require the provision of special control equipment.

For many types of operations, gas is a very efficient fuel. However, for other cost reasons many plants are located in rural or suburban areas that do not have a gas supply system. The cost of providing such a system for one large user would be prohibitive, and so plants must have gas storage facilities. For various reasons, the stored gas is usually pressurized fuel in the form of liquefied petroleum gas (LPG) or liquefied natural gas (LNG).

Building gas systems are valuable tools in any industrial society if installed and used properly and maintained with due regard to the explosion and fire potential. Misused, they can be the source of major disasters that could have been avoided by proper attention to a few fundamentals.

I. Properties of industrial gases used in buildings' utility systems.

Building Utility Systems

Gases used in utility systems vary widely in their composition, the pressures at which they are used, their explosive ranges, and their vapor densities. There are also other points of difference, but, insofar as the average fire prevention inspector is concerned, the factors listed above are of primary concern. It is also essential that he know the interrelationships between temperature, pressure, and volume.

In analyzing the potential fire hazard of a flammable gas that is being used or is proposed for use, it is usually best to begin with the basic gas laws of pressure, temperature, and volume and then consider the question of explosive range. The basic laws are as follows.

A. The volume occupied by a given mass of gas varies inversely with the absolute pressure, assuming a constant temperature. This means that $PV = $ a constant, or $P_1/P_2 = V_2/V_1$, or $P_1V_1 = P_2V_2$.

Note: Absolute pressure = gauge pressure + 14.7 psi at sea level.

B. The volume of a given mass of gas varies directly with the absolute temperature, assuming the pressure is kept constant. This means that $V/T = $ a constant, or $V_1/V_2 = T_1/T_2$. T_1 indicates a starting condition and T_2 indicates a changed condition.

Note: Absolute temperature (using Fahrenheit scale) = 459 + degrees above 0° F. or minus degrees below 0° F.

C. The working relationships among pressure, volume, and temperature may be expressed by combining the two above formulas to obtain the formula $T_1/T_2 = P_1V_1/P_2V_2$.

The following example illustrates how to use this formula. A 1,000-cu. ft. refrigerated tank stores propane at −70° F. and 800 psi. The tank ruptures when the ambient temperature is 100° F. When the gas reaches ambient temperature and pressure, what will be its volume?

$T_1 = 459° − 70° = 389°$. $T_2 = 459° + 100° = 559°$.
$P_1 = 800 + 14.7 = 814.7$ psi. $P_2 = 14.7$ psi.
$V_1 = 1,000$ cu. ft. $V_2 = ?$

By substituting numbers for symbols in the formula, we obtain the equation $389/559 = (814.7 \times 1,000)/14.7 V_2$. $V_2 = 79,642$ cu. ft.

Offhand it might not look like much of a problem to have a 10 ft. by 10 ft. by 10 ft. cube of gas expand to eighty times its original size. But before a reader takes such a situation lightly, he should look at the gas's explosive range (concentration at which explosion is likely), which for propane is 2.2% to 9.6% when mixed with air. Unlike solids, which retain their size and usually their shape, and liquids, which retain their size with little change, gases expand to fill the space available. In addition, they mix readily with other gases. If the situation cited above should result in a perfectly possible air-propane mix of 7% propane and 93% air, an explosive mixture of air and flammable gas would fill a volume of approximately 1,140,000 cu. ft. If the hazardous mix had an average depth of 10 ft., more than 2½ acres would be in jeopardy of an explosion.

It is important to remember, then, that gases are elastic and have no fixed volume except as arbitrarily confined and that they will mix readily with air and other gases. *The potential hazard area is not limited by the size of the container. It can go far beyond the original area.*

Because a gas can vary widely in volume, depending on pressure and temperature, and because a flammable gas can mix with air to form an explosive mixture, it is also vital to know the vapor density (relative density compared to air) of the gas or gases being used. On the vapor density are based important ventilation requirements and the locations of storage tanks and other storage facilities. Different types of ventilation systems are required for gases that are lighter than air and for those that are heavier than air. The two types of gases also require different kinds of storage facility arrangements. The following table lists explosive ranges and vapor densities for some of the more common gases used in utility systems.

Table 1

Gas	Explosive Limits in Air % by vol.		Vapor Density Air = 1.0
	Lower	Upper	
Butane	1.9	8.5	2.06
Methane	5.3	13.9	0.55
Propane	2.2	9.6	1.56
Natural Gas (High BTU)	4.7	14.5	0.62 to 0.72
Carburated Water (Normal)	6.4	37.7	0.64

II. Storage facilities for liquefied and pressurized gases. There are, of course, many storage facilities operated by utility companies, but these do not normally come within the purview of the average fire prevention officer except occasionally as a potential exposure. These storage facilities do not, in general, pose a major fire problem because they normally contain gas that is lighter than air, at a low pressure, and at the ambient temperature. In contrast, the liquefied petroleum and natural gases used in large quantities at many industrial plants and at some mercantile and institutional facilities are stored at high pressures and may also be refrigerated. It is with such facilities that fire prevention generalists should be concerned. For example a plant might have a storage tank 50 ft. in diameter and 20 ft. high filled with LNG refrigerated to a temperature of $-150°$ F. and held at a pressure of 500 psi. Outside temperature is $90°$ F. We can calculate the explosive potential of such gas by applying the basic formula $P_1/P_2 = T_1V_2/T_2V_1$.

$P_1 = 500 + 14.7 = 514.7$.
$P_2 = 14.7$ (atmospheric pressure).
$T_1 = 459 - 150 = 309$.
$T_2 = 459 + 90 = 549$.
$V_1 = 3.14 \times 25 \times 25 \times 20 = 39,250$ cu. ft.
Explosive range $= 5.3\%–13.9\%$. Use 7%.

By substituting numbers for symbols in the formula, we obtain the equation $514.7/14.7 = 309V_2/(549 \times 39,250)$. $V_2 = 24,000,000$ cu. ft. $\pm$ of gas. Assuming the gas mingles with air to form a mix that is 7% gas, by volume, we can calculate the volume of explosive gas-air mix as follows: $24,000,000/0.07 = 343$ million cu. ft. $\pm$ of a gas-air mix that is in the explosive range. Assume that the gas-air cloud is 30 ft. high, and it is theoretically possible to place 260 acres in danger of explosion. Even with an allowance for gas escape on the cloud fringes, 175 acres or more would still be at hazard. Such tanks should never be permitted in a settled area, and in any location require very careful attention to siting and other considerations. At the present time, storage tanks of this size are few and carefully regulated, but there is a trend towards tanks larger than those currently in common use. This trend should be watched so as to keep permissible tank sizes in line with local conditions and control potential hazards.

The National Fire Protection Association, the Factory Mutual System, and other nationally recognized fire safety authorities publish guidelines detailing desirable distances between various sizes of LPG and LNG storage tanks and important structures and property lines. Any one of these publications makes a desirable part of a good fire prevention library.

If stored gas is at ambient temperature, the vapor density of the gas is of great importance in siting tanks. Lighter-than-air gases that escape from a tank tend to rise, mixing with air in the process. Escaped heavier-than-air gases also mix with air, but they gradually settle to the ground level and follow downward slopes to low points. For this reason, LPG tanks of any size should be located where the ground slopes downward from buildings, highways, or other populated areas to an area where the gas can safely dissipate. Where ground conditions are not favorable, diversion or containment dikes, as appropriate, may be provided. Another possible safety precaution is to bury the tanks or mound them over.

LNG may be stored at ambient temperature or refrigerated. LNG at ambient temperature is lighter than air, but the vapor density of refrigerated LNG at release may be greater than the air density. Under such conditions, liberated LNG hugs the ground until it warms up and becomes lighter than air.

Many pressurized gas tanks are in the shape of horizontal cylinders. Regardless of ground slope, these tanks should never have the long axis pointed towards highways, important buildings, or other tanks. In the event of a tank failure, the tank might take off like a rocket. (In a number of tank head failures, this happened, and the tank traveled several hundred feet.)

Fortunately, gas storage tank failures are rare. However, because of the potential for enormous damage in the event of failure, gas tanks should always receive attention.

III. Vaporizers and regulators. Use of liquefied gases for building utility systems requires vaporization of the gases. In many instances, particularly where gas withdrawal rates are fairly uniform and moderate, the natural vaporization at the container location is adequate. For systems where the demand is high or fluctuates widely, vaporizers are often needed to speed up the conversion from liquid to gas. These vaporizers may be heated by steam, hot water, electricity, or even by gas burners. Preferably, the vaporizers should be located with other gas controls in a noncombustible building that is detached or against a blank wall of noncombustible or fire resistive construction.

Design of vaporizers is a specialized field. The fire prevention inspector should make certain that the installation was designed by qualified personnel and that it is being properly maintained.

Regulators are designed to assure the delivery of gas at a constant pressure regardless of variations on the supply side. For example, if a distribution system operates at a pressure of 30 psi and the plant uses gas at a pressure of 1 psi, a regulator is needed. Regulators are normally present in a vaporizer building and may also be present elsewhere. The devices require good maintenance because, if one fails, full gas pressure may come on a system not designed for it. (For example, the author is aware of an instance where regulator failure at a vaporizing plant resulted in full pressure on a distribution system serving the homes of 6,000 people and the attendant business activities. De-

Building Utility Systems 87

spite prompt operation of shutoff valves, some devices failed, and there was a major fire. Also, service to shops, stores, and homes was interrupted for a considerable period.)

IV. Basic controls. A building may have few or many controls, depending on its size and the extent of its gas system, and each control has a definite purpose. For the system as a whole, however, one type of control is of primary concern. Each supply line entering the building should have an accessible and properly maintained shutoff valve. A building fire may involve the main gas lines whether or not it originates in gas piping. For this reason, it must be possible to shut off the gas supply from a safe location outside the building.

V. Gas piping systems. All gas piping systems have certain features in common, but they differ in some respects depending on whether the gas in use is heavier or lighter than air. However, it is possible for a system designed for LPG to be converted to LNG or vice versa. There may also be structural building changes that add or remove crawl spaces or affect the system in other ways. Assuming that a gas piping system has been properly designed and installed, it must still be maintained. For these and other reasons, fire prevention personnel should be aware of certain basic installation and maintenance rules.

  A. Heavier-than-air gases.

    1. Piping should be above ground whether outdoors or indoors. If outdoor piping must be buried, it should rise above ground before entering the building.

    2. Indoor piping should not be located in crawl spaces, basements, or other inaccessible locations.

    3. Pipe sleeves in building walls should be sealed gas tight. (See Figure 8.)

    4. Piping should be protected against mechanical injury as necessary.

    5. Where gas pressures in excess of 30 psi are used, piping should be encased in larger-diameter pipe that is vented to the outdoors.

    6. Each main supply line should have an independent shutoff valve. These valves are in addition to the main shutoff outside the building.

    7. Whenever it is necessary to shut down the supply line for alteration and/or repairs, not only should the main control valve be shut, but open-end pipes should be capped. A number of disastrous explosions have resulted when a leaking control valve fed gas into open-end piping.

    8. All piping should be protected against corrosion and be regu-

Fig. 8. LPG line through building wall

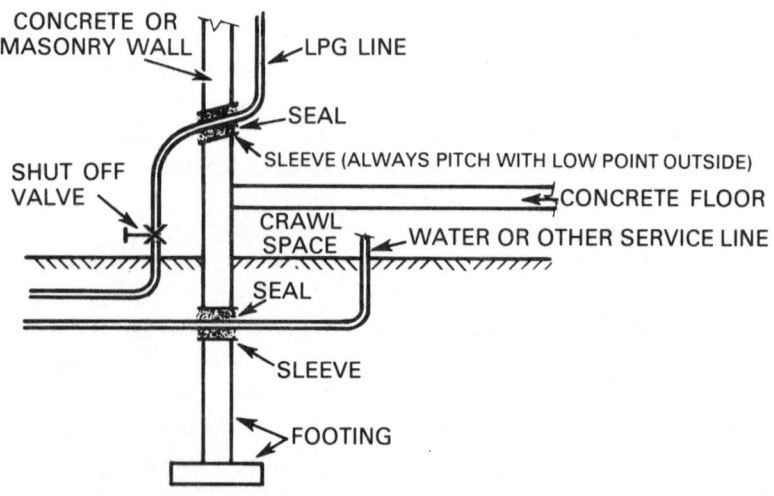

larly inspected by qualified personnel; all necessary maintenance must be performed.

B. Lighter-than-air gases.

1. Piping outdoors may be either above or below ground. Indoor piping, however, should not be buried.

2. Certain lighter-than-air gases contain moisture that may condense and freeze. Piping handling these gases in unheated areas should pitch back to drain at properly located condensate outlets. It is particularly important to watch this point when a heated building is converted to an unheated facility.

3. Piping may enter a building below ground, but it should not run under a foundation wall or footing. Openings into the building should be sealed gas tight.

4. If piping is installed in a crawl space or unfrequented basement, the area should have fresh air inlets.

5. Floor openings for riser piping should have gas-tight seals (see Figure 9). (A few years ago, uninspected or poorly inspected piping in an apartment building crawl space was a major factor in a serious fire. The crawl space lacked ventilation, and the floor openings lacked seals. The leaking gas filled the crawl space and

Building Utility Systems 89

Fig. 9. Venting crawl space: lighter-than-air gases

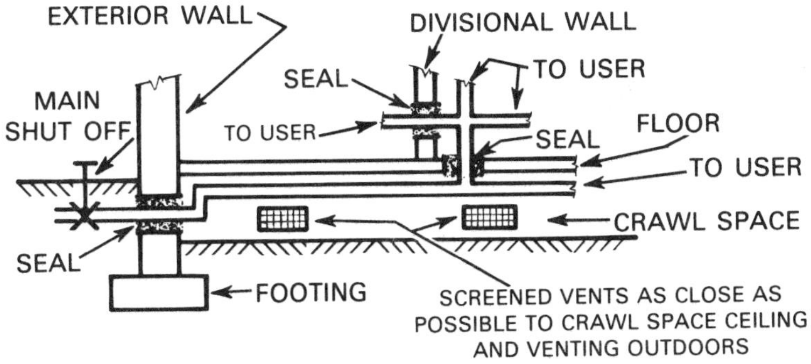

moved up into the building until it found a source of ignition. The resulting explosion killed or injured a number of people and destroyed the building.)

6. Piping should be protected against mechanical injury and corrosion and be properly maintained.

C. Conversion programs. Whenever a gas system is to be converted to handle a different type of gas, the work should only be done under the close supervision of qualified personnel. Such a conversion may involve replacing controls, burners, or even some piping.

D. General. Regardless of the type of system in use or proposed for use, there is always the possibility that systems may be shut off or turned on for any of a number of reasons. Fire prevention personnel have a responsibility to see that preplanning provides for this work to be done by experts in the field.

Heating Systems

Contrary to popular belief, heating systems are a major source of fires. Besides causing direct losses, heating system fires occasionally are responsible for long-range indirect losses that may be far more costly than the immediate destruction. The long-range impact is, unfortunately, often overlooked.

There are many types and sizes of heating boilers, many methods of firing, and many types of heat distribution systems. However, fire prevention personnel should be concerned with basic fire safety questions, not detailed design criteria. Although there are other areas of interest, loss statistics and reports that the author has seen indicate that the great majority of heating system fires occur because of failures in seven areas: (1) fuel safety, (2) firing procedures, (3) essential fire cutoffs, (4) control of combustibles, (5) chimneys and stacks, (6) testing and maintenance, and (7) spare capacity.

I. Fuel safety.
  A. Coal.
    1. Storage. Bituminous coal is susceptible to spontaneous heating, particularly if it is fresh from the mines. Such coal also gives off flammable gases in varying amounts. Control of these fire potentials is largely obtained by observing the following safety measures.
      a. Avoid outside storage over pipe trenches or manholes or around other structures that may permit excess air movement through the coal.
      b. Avoid contact with outside sources of heat, such as steam mains.
      c. Use the roll-pack storage method to prevent fine coal from sifting out from the larger pieces. This storage method inhibits air movements through the coal.
      d. Vent inside storage bins to remove flammable gases given off by the coal.
    2. Pulverized coal. Use of pulverized coal presents a much greater explosion potential than any other type of coal use. Explosions can result if a mix of coal dust and air or flammable vapor and air comes in contact with an ignition source. Most recorded explosions have occurred in pulverizers, fireboxes, or cyclone collectors. For example, pulverized coal, for any one of a number of reasons, may hang up somewhere in the system. The coal must be dislodged so that it will start flowing again. This job wrongly done may result in a dust cloud that would explode violently in the presence of an ignition source. For another example, coal in a collector or storage bin might spontaneously heat up and generate flammable gases. Unless means are available to vent these gases, the formation of an explosive mix of flammable gas and air is quite possible. Finally, because pulverized coal explosions tend to be quite violent, it is essential that the transmission system from the pulverizers to the fire box be provided with explosion vents to the outdoors.
  B. Oil. Insofar as fire safety is concerned, storage and piping arrangements for oil are similar to those for other flammable liquids in the same hazard range. Such arrangements are detailed in Chapter 9, Flammable Liquids and Gases.

Building Utility Systems

C. Gas. Storage and piping arrangements for gas used for heating systems are as previously noted in this chapter. Additional information is included in Chapter 9, Flammable Liquids and Gases.

II. Safety of firing procedures. Heating boilers may be fired by manual, semiautomatic, or automatic means. The method of firing and the type of fuel used determine the nature of the controls necessary for safe operation. There are literally hundreds of kinds of control devices and many possible combinations of controls, but these are primarily of interest to the designer or the specialist. Fire prevention personnel should be primarily concerned with those areas where the record shows mistakes have resulted in serious fires and/or explosions.

A. Improper lighting-off procedures. The principal deficiencies in this area are as follows:

1. Repeated attempts to light off without allowing sufficient time between attempts for vapor concentrations to leave the firebox.
2. Bypassing or tampering with controls.
3. Firing off the firebox walls. This sometimes happens when an attendant is in too much of a hurry to bring a hot boiler back on the line.

B. Attendance. Virtually all building and fire codes require continuous attendance on boilers over a specified size or operating over a specified pressure. (For example, if an inspector walks into a boiler room and finds a 400-h.p., manually fired boiler operating at 75 psi, he needs to assure himself or herself that the plant is providing attendance in the boiler room at all times. Failure to do so is flirting with disaster.)

C. Inadequate or improper controls. Some of the common control deficiencies are as follows:

1. Resetting controls so that trial for ignition periods is too long.
2. Wrong combustion controls. (The writer remembers an oil-filled boiler of considerable size and pressure that blew because it had a relatively slow-acting stack switch instead of the fast-acting combustion control it should have had. Eighty feet of wall went down. Stack switches are fine when used for their designed purpose but not when used in the wrong place.)
3. Unapproved equipment. Combustion controls, oil valves, gas valves, and other equipment used for burner controls should be UL or FM approved. Occasionally, inspectors will find that unapproved units have been installed as replacements for the original units.

4. Interlocks. An inspector may occasionally find missing or bypassed interlocks between the fuel system, air supply, and pilot supervision on automatic units.

5. Equipment malfunctions. Most of the potential for malfunctioning equipment will be overcome by proper maintenance and adherence to manufacturers' instructions on operating and adjusting procedures.

6. Special pilot precautions. When pulverized coal is used as a fuel, there should be a large, stable pilot flame, either gas or oil. Manual means for igniting a pulverized coal boiler should be avoided.

III. Adequacy of fire cutoffs.

A. Boiler rooms. In general, boiler installations of any size are separated from adjoining areas by walls whose fire resistance is specified by the building and fire codes in use. There are, however, certain other factors that need to be considered and perhaps emphasized a little more than they usually are. For example:

1. Exits. A single means of egress may be permissible in a small or moderately sized boiler room, particularly in a ground level room that has windows. However, larger rooms usually require two exits to assure safety to life. These exits should be as remote from each other as possible, and one should preferably lead outdoors.

2. Occupancy in adjoining areas. Many boiler room floors are at the same level as adjoining floors or below them. Ordinarily this is no problem. But if occupancy changes bring in flammable liquids and there is any possibility that a vapor hazard may develop from a flammable liquids leak, it may be desirable to install a ramped curb at the entrance to the boiler room.

B. Boiler houses. Many heating and power boilers are installed in separate buildings. Such buildings should be constructed of noncombustible or fire resistive material as appropriate. If combustible construction is used, automatic sprinkler protection should be provided.

C. Horizontal and vertical fire cutoffs. The system used to distribute heat from the boiler plant to the areas heated may be steam, hot water, or hot air. It is probable that the piping or ductwork involved will pierce many of the fire walls or fire resistive floors that a building may have. Where this occurs, it is vital that the fire integrity of the cutoff be maintained. In most piped systems, clear spaces between piping and pipe sleeves should be packed with non-

Building Utility Systems

combustible material. Maintaining fire integrity in duct penetrations was covered previously in this chapter under the heading Air Handling Systems.

IV. Control of combustibles. While the need for keeping boiler rooms and boiler houses free from combustibles is generally well recognized and needs no comment, there are areas that are frequently overlooked, as outlined below.

A. Combustible floors and partitions. Steam pipes often operate at temperatures well above 212° F., and hot water pipes may also reach such temperatures. Such hot pipes may carbonize abutting wood so that it turns to charcoal, and the charcoal may then ignite. Clear spaces must be maintained between such piping and combustible floors or walls.

B. Drying racks. Such highly combustible materials as paper and cotton cloth should not be placed on radiators or pipes to dry. In addition, spaces behind radiators or pipes should be kept free from combustible material. (About ten years ago, the author investigated a serious chapel fire that originated in a chaplain's office. It was conclusively proven that the fire originated in an accumulation of combustibles that had dropped down behind the radiator and gradually carbonized. Similar carbonization in various stages on the road to ignition was noted in combustibles that had fallen behind other radiators in the unburned portion of the building.)

V. Testing and maintenance procedures. Whenever a heating boiler of any size is installed, it is, or should be, standard practice to provide operating personnel with testing and maintenance instructions. Such instructions should be posted or otherwise be available in the boiler room. Along with them should be a record of the actual tests and maintenance done, by date and by whom. Records should note what deficiencies, if any, were found and what corrective actions were taken.

In addition to normal testing and maintenance activities, boiler room procedures should provide for the instruction of new personnel and for such additional instruction for current personnel as might be required by changes in equipment or operational requirements.

During inspections, fire prevention personnel should, when feasible, witness tests of the combustion safeguards provided and the ensuing lighting-up procedures. Information on permissible times for the operation of safeguards and for the trial for ignition period is readily available.

VI. Space capacity. In making a fire prevention inspection, one of the easiest things to overlook is the potential for heavy indirect losses from fires that by no stretch of the imagination could be conceived of as resulting in more than a few thousand dollars in direct property damage. Nowhere is this error more likely to be made than in boiler rooms.

Assume, for example, that Facility A has a well-arranged boiler room with brick walls, concrete ceiling and floors, and adequate exits. In the room is one fair-sized, oil-fired boiler that does not have the proper combustion safeguards and that has somewhat substandard maintenance. Looking it over, the inspector decides there is a slight to moderate possibility of a burner backlash that might result in a five- to seven-thousand-dollar loss with damage confined to the boiler room. Recommendations are made for improvements but without any sense of urgency. How right is this approach?

A. The boiler is a standby unit that is used only occasionally. In this case, the decision is correct.

B. The boiler is the only important source of heat available for a 250-bed hospital in a cold climate. During cold weather, a fire or explosion in this boiler could easily create a major emergency. The decision in this case is incorrect. The need to assure the hospital's capability to care for patients is urgent.

C. The boiler is one of five, of which only four are needed to carry plant peak loads. The decision in this case is correct.

D. The boiler is a single unit and supplies heat for critical plant operating areas. The decision in this case is incorrect. Loss of heat in these areas could easily result in a plant shutdown of weeks or months.

In an outside boiler house, another factor may interrupt heating service. Assume the boiler is properly equipped and maintained but is located in an unsprinklered, combustible boiler house. It is possible to burn off a wood roof without serious fire damage to the boilers, but how are they to be operated in the open at below-zero temperatures?

In summing up the question of spare capacity, it should be emphasized that fire prevention inspections of boiler rooms or boiler houses should always include consideration of the possibility of serious indirect losses from boiler fires and/or explosions. The writer recalls participating in two fire loss investigations of this nature. In one instance, a burner backlash resulted in $2,500 worth of direct property damage and $150,000 worth of indirect losses resulting from a plant shutdown. In the second instance, direct losses amounted to about $3,000, and indirect losses totalled about $125,000.

VII. Chimneys and stacks. Defective chimneys are an important source of fires and are responsible for substantial direct and indirect fire losses. Chimneys and stacks range from a few feet to several hundred feet in height and from a few inches to a number of feet in diameter. In addition, they differ in the materials used in construction. However, only a few readily observable defects in equipment or procedure are responsible for most of the chimney and stack fires, and proper plant maintenance programs will eliminate these defects. Of primary importance are the following.

A. Inspections. All important chimneys and stacks should be inspected at regular intervals. An inspector should check such features as the condition of the firebrick lining, the capability of lightning rods to provide proper resistance to ground, the need for internal cleaning, the condition of mortar in brick stacks, clearance from combustibles, and corrosion of metal stacks. Findings, together with recommendations for correction, should be a matter of record at the facility.

B. Lightning protection. Resistance to ground is a key factor in lightning protection. Since ground resistance does not remain constant, it should be checked annually. Additional grounds should be provided if the tests indicate them to be necessary. It should also be noted that, if a brick stack of a given height requires lightning protection, any nearby metal stack of similar size should also be checked for adequate grounding. (Metal stacks should be protected, not with lightning rods, but by grounding their bases.)

C. Spark arrestors. Where solid fuels or combustible residues are burned near combustible roofs, lumber storage, or other combustibles, spark arrestors should be provided for chimneys and stacks other than those of unusual height.

D. Clearances. All chimneys and stacks require clearances from combustible construction. Required clearances will vary from approximately 2 inches for masonry stacks to a number of inches for metal stacks. Clearance requirements for metal stacks vary according to whether the appliances served generate low, medium, or high heat.

E. Mortar. During each inspection, brick stacks should be checked for mortar deterioration. Crumbling mortar should be promptly replaced. Bad joints may not only threaten the structure of a stack, but also allow excessive heat to escape.

F. Unburned gases. In many instances, the gases in large chimneys or stacks are monitored for the presence of carbon dioxide, carbon monoxide, or other kinds of gases. Where stacks are large enough to warrant this type of monitoring, it is vital to carefully check the amount of unburned flammable gases in the stack. (An explosion in a 400-ft. stack could very well shut down a plant for months.)

Keeping stacks operational and eliminaing them as ignition sources is a very large step toward maintaining continuity of operations and decreasing fire losses.

chapter 6
Fire Protection Systems

A little over one hundred years ago, there were no installed fire suppression systems. Efforts to decrease the number and severity of fires consisted chiefly of improving construction standards, occupancy safeguards, and firefighting capabilities. These advances did result in some decreases in fire losses, but the hoped-for sharp drop did not materialize. Something was still missing from the fire safety picture.

Today, many people are familiar with the analogy between a fire resistive building loaded with combustibles and a furnace loaded with fuel. Both structures can withstand, at least in most cases, a total burnout. For the furnace, this is fine, but for a factory, it spells disaster. To carry the analogy a step further, it is possible to control the rate at which fuel burns in a furnace or even to drop the fire completely. It should, therefore, be possible to develop means to control or limit the spread of fire in a building and ultimately to extinguish the fire. Some such line of reasoning was in the minds of industrialists and others in the years immediately following the Civil War. They tried various experiments along this line and, in 1874, the automatic sprinkler system was born.

As with most new developments, the acceptance of automatic sprinklers was slow at first, but such systems soon proved their worth. Once the use of automatic sprinklers became widespread, the number of serious fire losses dropped sharply. This success spurred the development of ever more efficient sprinklers and of new types of systems to meet particular needs.

Following the success of sprinkler systems, some farseeing individuals turned their attention to the development of fire control means for those occupancies where sprinkler protection is not feasible. Such occupancies include those that are unusually susceptible to water damage or that use chemicals that react violently with water. Research in this direction led to such present-day special systems as those using carbon dioxide, dry chemicals, foam, or halon. Used in the right places, these systems provide valuable protection. They should not, however, be considered as a substitute for sprinkler systems, which are the backbone of installed fire suppression systems.

Through the years, people recognized that installed fire suppression systems could do even more than directly control and extinguish fires. For example, these systems can be made to transmit fire alarms, close or open doors and windows as appropriate, start fire pumps, control air flows through air conditioning and ventilating systems, stop industrial processes, or open wall panels.

So that the subject of fire suppression systems will not appear to be more complicated than it really is, it should be emphasized that, except for a few specialists, fire prevention personnel are not designers. What they need is knowledge in a few specific areas, such as those listed below.

1. Basic capabilities of each type of system, what it can and cannot do.
2. Whether or not a system needs to be arranged to perform tasks in addition to its normal fire control and extinguishing functions.
3. How to read working (not design) drawings in order to make field checks of a new installation.
4. How to make operating tests and how often to repeat them. In most instances, the inspector witnesses tests made by others.
5. What the appropriate standards of maintenance are.

Once he has learned these things, a fire prevention inspector is in a position to advise on the acceptability of existing systems, the need for changes in existing systems, or the need for new installations.

Automatic Sprinklers

Before discussing the several types of automatic sprinkler systems, it is appropriate to note characteristics shared by all sprinkler systems, to discuss where sprinkler systems should be installed, and to clear up some common misconceptions about sprinkler system operation.

I. Causes of sprinkler failure. Even with today's inflation, the fire loss record shows that the average loss in a fully sprinklered building is

under $2,500. The record also shows that, when a large fire loss does occur in a sprinklered building, the loss is usually due to one of the following causes.
  A. Main control valve closed.
  B. Inadequate water supply.
  C. Obstructed piping.
  D. System not designed for type of hazard involved.
  E. Shielded fire out of reach of sprinkler discharge.
  F. Explosion that shattered sprinkler piping.

Note: All of these potentials for large fire losses can be eliminated by following maintenance and testing procedures for sprinkler systems and by establishing adequate occupancy safeguards.

II. Where to install sprinklers. A few lives have been lost in sprinklered buildings, but the number of casualties has been far less than in similar buildings that were unprotected. For the most part, loss of life in properly protected buildings has been due to explosions or to initial flash fires. Sprinklers save lives as well as property.

Automatic sprinklers should be installed wherever building construction is combustible, where there is appreciable continuity of combustible contents regardless of building construction, where a vital operation involves combustibles, and for the protection of life as appropriate. In applying these guidelines, it is essential that the following factors be taken into account:
  A. Values. At the present time, as a rough rule of thumb, when values exposed to a single fire potential approach $250,000, the cost of sprinkler protection can usually be written off from insurance savings alone.
  B. Vital operations. Loss of a vital operation may inflict indirect losses far in excess of actual fire damage. (For example, loss of a small, unsprinklered, combustible chlorination plant could well cripple a small community water supply system.)
  C. Life safety. Sprinkler installation for life safety depends on both the building and its occupancy. (For example, a low-value, two-story combustible, barracks-type building is converted from sleeping quarters for physically able adult males to quarters for aged and infirm people. Sprinkler protection should be required before any such conversion is made.)

III. Misconceptions. There are two common misconceptions about sprinkler system operations that need correction.
  A. Many people believe that, in the event of fire, all the sprinkler heads in the fire area operate. Except for deluge sprinkler systems,

which are special purpose systems, this is not true. The record shows that 76 percent of all fires in sprinkler protected buildings are controlled by five heads or less and 95 percent by twenty-five heads or less. In other words, three-quarters of the fires were controlled or extinguished by a sprinkler water discharge of 100 gpm or less, and 95 percent of the fires by 500 gpm or less.

B. Many people think that the installation of sprinklers over high-voltage electrical equipment is hazardous to personnel. This is false, as has been proven many times.

IV. Types of systems and their uses.
  A. Wet pipe systems.
    1. Characteristics. Wet pipe systems account for 75 percent of all sprinkler systems currently in use. Basic reasons for the predominance of these systems are as follows:
      a. Speed of operation. Because a wet pipe system has water instantly available at the sprinkler heads, it can respond more quickly than a dry sprinkler can. As a result, fewer heads operate and less water damage occurs.
      b. Speed of system restoration. Normally all that is required to restore a wet pipe system to service after a fire is to replace operated or weakened heads, reopen control valves, and make a two-in. drain test. Purpose of the drain test is to make certain there are no obstructions in the supply piping.
      c. Ease of maintenance. Basic devices are few in number and easy to maintain.
      d. Extensions. In most instances, wet pipe systems can be extended considerably without adding new sprinkler risers. Basic limitations on proposed extensions are adequacy of available pressures, adequacy of distribution pipe sizes to support the extension, and code limitations on the size of fire area that can be protected by one system.
      e. Sprinkler head position. Sprinkler heads can be installed upright or pendent as desired, generally with no effect on cost. However, dry pendent heads drop-nippled into such unheated areas as show windows can run costs up.
      f. Small hose. Subject to insurance company approval, small hose connections (for fire protection purposes only) are permissible on wet pipe systems.
    2. False alarms. False alarms are not, strictly speaking, a fire prevention matter. However, frequent false alarms on a system may cause the system to be turned off. From this standpoint, false

alarm situations are of interest to fire prevention personnel. Where risers are supplied from a system subject to pressure surges of thirty to forty seconds or longer, false alarms may become a problem. Some of the methods of eliminating false alarms are as follows:

    a. On divided seat-ring alarm valves, install a ¼-in. bypass with check valve around the alarm valve clapper.

    b. On pilot valve alarm valves, increase the pilot valve clearance.

    c. Maintain 20 psi excess pressure above the alarm valve.

    d. Install a second retard chamber between the alarm valve and the alarm itself.

3. Unheated areas. Wet pipe system piping may pass through small unheated areas or be used to supply a few heads in an unheated area. The following procedures are in order for such cases.

    a. Protect exposed piping with a heated enclosure or with insulation, as warranted. It may be necessary to install a steam tracer line.

    b. Install nonfreeze or small dry pipe systems in unheated areas. Be sure connections with the main system are in heated areas.

4. Cold weather valves. In many cases, one or two heads are in an unheated area and are controlled by a valve that is shut in winter. Care should be taken to make certain that these valves are reopened when freezing conditions are over.

5. Hazard recognition. For purposes of automatic sprinkler protection, occupancies are divided into three classes: light hazard, ordinary hazard, and special (or extra) hazard. With rare exceptions, original installations are designed to fit the hazard involved. However, the inspector needs to be alert to occupancy changes. If the hazard rating remains the same or is lower, there is no problem. On the other hand, converting an ordinary hazard occupancy to an extra hazard occupancy may well require major changes in the sprinkler system to assure adequate protection.

B. Dry pipe systems.

1. Characteristics. Dry pipe systems are used in unheated buildings and other areas where water in the piping may freeze and it is not practical to provide heat. System piping is filled with pressurized air, which holds back water at a dry pipe valve in a heated location. Operation of one or more heads releases the air pressure, the dry pipe valve trips, and water enters the system. Because of

the time required for water to reach a fire, dry pipe systems are somewhat less efficient than wet systems. As a result, other things being equal, more heads will operate in a fire situation. Other differences from wet systems include, but are not limited to, the following.

a. System restoration. To restore a dry pipe system to service after a fire, it is necessary, not only to perform the tasks listed for wet systems, but also to reset the dry pipe valve and restore air pressure in the piping. Of course, the added tasks take additional time.

b. Maintenance. The basic devices in dry systems are more numerous and complex than those in wet systems and require more maintenance. Besides such devices as dry pipe valves and control valves, dry pipe systems have air compressors, low point drains, heating facilities for valve houses, and other equipment.

c. Extensions. Extensions to dry pipe systems are limited by the air capacity of the piping as well as by hydrostatic pressures. Where there are dry pipe valves only, the permissible system capacity is 500 gallons. If accelerators or exhausters are added, the permissible capacity is increased to 750 gallons. If the system has a normal riser with limited horizontal run, it is satisfactory to estimate system capacity on the basis of $1\frac{1}{2}$ gallons per head.

d. Hazard rating. Except where it is not practical to keep temperatures above freezing, as in grain elevators, dry pipe sprinkler systems should not be used for extra hazard situations because of their being slower acting than wet systems.

e. Water columning. Water, whether from leakage past the dry pipe value or from condensation of water vapor in system air, should not be allowed to column in the riser above the dry valve. The water may of course freeze, and, in addition, a relatively low head of water prevents many dry pipe valves from tripping when heads operate and air pressure is released.

f. Conversions. Quite often, an occupancy change eliminates requirements for heat. When this occurs and conversion of a system from wet to dry is proposed, particular attention must be paid to the pitch of the piping. All dry piping should be pitched back to a main drain or to appropriately located low-point drains.

g. Vertical versus horizontal systems. In wet systems with multiple risers, the differences between vertical and horizontal de-

Fire Protection Systems

sign are quite minor. In dry pipe systems, the differences can be of major importance. Assume a three-story, brick-joisted building with 275 heads per story. If each story has its own system, a small fire in one story will operate only one system. But if three systems are installed vertically, the same fire might operate all three systems. Should temperatures be appreciably below freezing, nonfire floors could suffer heavy damage. (See Figure 10.)

C. Preaction systems.

1. Uses. Preaction systems are used where occupancy conditions are such that—

   a. it is desirable to have an alarm in advance of sprinkler operation;

   b. it is especially important to prevent the accidental discharge of water because of the potential for serious water damage in the event of broken piping or damaged heads;

   c. it is important to be able to work on a portion of a system while keeping the bulk of the system in service; or

   d. speed of a wet system is desired, but other conditions require dry piping.

2. Operating principles. Operation of a preaction system requires

Fig. 10. Dry pipe system arrangements

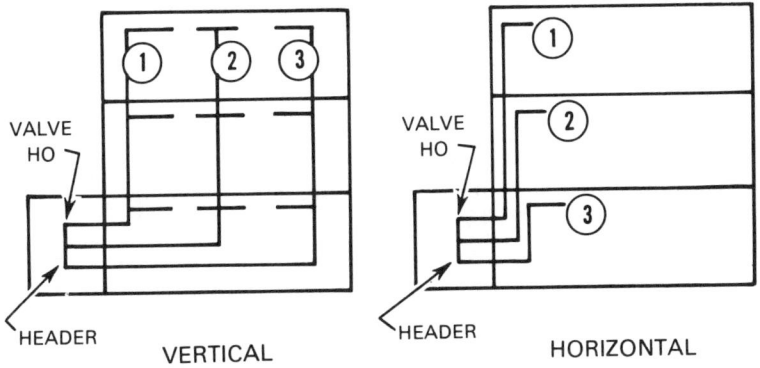

the presence of a system of heat-actuated devices that are set to operate before the sprinklers do. Operation of the heat detector trips the preaction control valve and allows water to reach the sprinkler heads almost as soon as they open. It should be noted that operation of a sprinkler, or disconnection or breakage of a pipe, will not trip the control, and so no water will enter the system as a result of these events alone.

3. Maintenance. Maintaining a preaction system is considerably more complicated than maintaining a standard wet or dry system. Personnel responsible for a preaction system should receive special training.

D. Deluge systems.

1. Operation. Deluge systems are designed primarily to wet down an entire floor area or hazardous operation. They operate by admitting water to sprinkler heads that are open at all times. Water for the sprinklers on each system is held back by a deluge valve. The valve is actuated by a heat response system installed throughout the sprinklered areas. Airplane hangars, rocket propellant handling areas, and pyroxylin plants are typical of locations where deluge systems are desirable.

2. Special concerns.

a. In hangars where large aircraft are housed, it is desirable that systems discharge foam as well as water. Water discharged from heads will not get under the airplane wings in a form that will control a flammable liquids fire.

b. Because deluge systems discharge a great deal of water, a fire prevention officer must always consider the possible need for special drainage. In some instances, there should be trapped floor drains leading to a safe location.

c. In large assembly plants, double lines of open sprinkler heads on deluge systems may be installed over wide aisles as a substitute for fire walls. Where this is done, the water supply for the deluge systems should be independent of the water supply for the building sprinkler systems.

d. There are three distinct types of deluge systems. The three types have different maintenance requirements and require various refinements to assure satisfactory operation. Therefore, it is usually desirable to select one type and follow through on that for new systems. Where a large staff of mechanics is available, use of more than one type is acceptable.

Fire Protection Systems

E. Exposure protection systems. Where there is a serious exposure problem, it may well be desirable to provide sprinkler protection over windows, under wooden eaves, or in other critical locations. Such systems may have open or closed heads and are normally supplied by connections to building sprinkler systems. Special concerns are as follows:

1. Plant operating hours. If a plant operates continuously or has watchman service during idle periods, open head systems are the most economical and the easiest to maintain. However, outdoor open heads may be blocked by debris or mud wasps' nests, and such blockage must be watched for.

2. If closed heads are used where there are no overhangs to trap heat, each head should be provided with some form of heat collector to assure operation.

V. Testing procedures.

A. Hydrostatic tests. All sprinkler systems must be hydrostatically tested at the time of installation and whenever major extensions or alterations are made. Large systems or extensions are usually tested one or more sections at a time. Such testing requires that blank gaskets or flanges be temporarily installed in the line. Each blank is identified by a lug or lip and must be removed following testing. It is important that each blank be accounted for. The author has seen a number of systems where a blank was overlooked. A sprinkler system is useless in limiting fire spread if water cannot reach the heads because of a blank.

B. Tests of 2-in. drains. During each fire prevention visit, all 2-in. sprinkler riser drains should be tested if possible. Such tests provide proof that the sprinkler control valve is open and indicate the condition of the supply main. If the water supply is adequate, opening the drain will normally cause a moderate pressure drop which will be recovered promptly when the valve is closed. Either a substantial change from previously recorded results or a prolonged recovery period provides a clear indication of trouble that should be located and corrected.

C. Air pressure tests. Every new dry pipe system is given an air pressure test. There are prescribed limits on the permissible air pressure drop over a twenty-four-hour period. Fire prevention personnel seldom get involved in these tests except to make certain they have been satisfactorily completed. An inspector should, however, be concerned when any abnormal amount of air leakage occurs. Such pres-

sure drops should be checked out and the leaks eliminated. Two common methods of locating air leaks are as follows.
  1. Soap joints and look for air bubbles.
  2. Introduce essence of peppermint or another strong-smelling substance into the air supply and note where odor appears on the system.
 D. Alarm tests. Water flow alarm tests should be made for each sprinkler riser during each inspection, as noted below.
  1. Wet systems should be tested by opening the inspectors' test connections near the end of the system. Alarm should ring in from three to five minutes. A regularly scheduled test may be postponed when a water-motor gong is involved and there is danger of freezing the gong, or when water discharged outside the building may form ice in a public area.
  2. Dry systems should be tested from the alarm by-pass except when dry pipe valves are trip-tested, in which case the inspectors' test connection should be used.
  3. On preaction and deluge systems there is normally a heat-responsive device used for testing alarms. Procedures differ according to the type of system in use.
  4. Many systems transmit alarms to fire departments or central stations. Make certain the alarm-receiving organizaion has advance notice of tests of such equipment.
 E. General.
  1. Depending upon what a sprinkler system is designed to do, other tests may be necessary, and these should be made as appropriate.
  2. All tests should be made by personnel assigned such responsibilities by the building owners. The inspector's functions are to observe and record and to make recommendations when necessary.
VI. Essential maintenance for sprinkler heads, valves, hangers, and allied equipment. Every plant, institution, or other facility having sprinkler systems and/or other fire detection and protection equipment should make provision for inspections of such equipment, preferably on a weekly basis. The plant should keep records showing tests made, conditions observed, recommendations given, and corrective actions taken. These records should be available for review by visiting fire prevention inspectors. Except in exceptionally clean occupancies, it is virtually impossible to make a good inspection without getting the record form dirty. Judging from past experience, the author places

little reliance on clean forms that no doubt have been filled out in the office after an inspection.

A. Sprinkler heads.

1. Overheating. Automatic sprinklers that are subjected to temperatures in excess of the safe maximum for their type may operate when there is no fire. These heads should be replaced by heads of proper temperature rating. Heads near a fire may also become overheated. Such heads should be examined and replaced as necessary.

2. Corrosion. Heads located in a corrosive atmosphere may become so corroded that they are inoperative.

3. Loading. In some occupancies, heads may become so loaded with spray or other residue that they become inoperative.

4. Paint. Sprinkler heads that have been painted should be replaced.

5. If there is any doubt that heads will operate properly, representative samples should be sent to a recognized laboratory for reliability tests.

6. Heads may also be damaged by freezing, mechanical injury, or other causes, but these are not very common.

B. Valves.

1. General. The following comments are applicable to all types of systems.

   a. Accessibility. All main control valves should if possible be located outside the fire areas whose systems they control. If this cannot be done, valves should be so located that they are readily accessible from outside the building. More than one building has been lost because fire got between fire personnel and a closed valve.

   b. Operation. To ensure that valves can be operated, each valve should be run completely up and down at least once a year. The writer has seen several valves that could not be moved either way because they had not been operated in years.

   c. Paint. Valve stems should never be painted.

   d. Valve direction. Preferably all valves at any plant should turn in one direction—that is, they should all be either right-hand or left-hand valves. During a fire not too many years ago, an employee inadvertently closed instead of opened two left-hand yard valves in an otherwise right-hand system. The result was a $15 million loss. To prevent this sort of event, all inside screw and yoke valves and other valves whose operation is not plainly evident should be clearly marked in some way to indicate the direction to open.

e. Tests. During each visit, try each valve to make sure it is wide open. Do not rely on post indicator targets and the like: they can slip. Also, do not rely too much on plant inspections. Dust accumulations on a valve wheel generally indicate plant personnel have not been trying valves.

C. Hangers. Inspectors should make sure there are no missing or loose hangers, a situation that may lead to a variety of problems. For example, a lack of hangers in a dry system might cause the pipe to sag. As a result, condensate may collect and possibly freeze. Such ice plugs have effectively cut off portions of a system, creating non-sprinklered area.

D. Dry pipe valves.

1. Tests. Valves should be trip-tested annually and repaired or serviced as necessary. Where several valves occupy a common header, reset each tripped valve before tripping another to avoid possible damage from water hammer.

2. Temperature. Temperatures in valve houses should be checked daily during freezing weather.

E. Allied equipment. As noted earlier, sprinkler systems may be designed to do many things, and such capabilities should be checked as appropriate. Information on these arrangements is in any standard reference textbook on fire protection systems.

VII. Flushing methods. If sprinklers are to perform their fire prevention function of limiting fire spread, the flow of water to them must be unobstructed. Obstructions can be removed by flushing yard systems, yard connections to sprinkler systems, and/or the sprinkler systems themselves as necessary.

A. Yard systems. New yard systems, additions to existing systems, or existing systems should be flushed through hydrants and blowoff valves. During flushing operations, valves controlling sprinkler risers should be closed to prevent obstructions from entering the systems.

B. Yard connections to sprinkler systems. New sprinkler risers are not connected to the yard system until yard connections are flushed. One way of doing this is shown in Figure 11. The simplest way to flush an existing system is usually to remove the clappers in the fire department connection and flush through the siamese.

C. Sprinkler systems. There are several methods of flushing, the choice being dependent on the kind of conditions encountered. This flushing should only be done by personnel who have been fully in-

Fig. 11. Flushing connection to riser

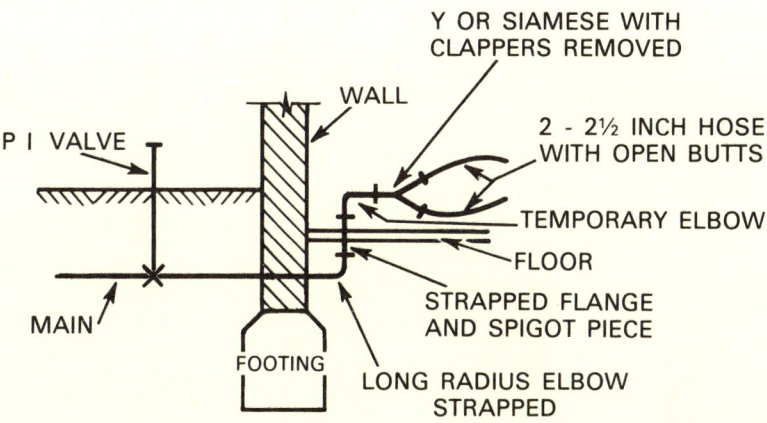

structed in the work. (Note: The fire prevention inspector does not do the flushing, but he or she is responsible to make certain that necessary flushing is done.)

VIII. Judging the adequacy of the system for the hazard protected.

A. Standards. *NFPA Standard 13* and other standard sprinkler rule books set up occupancy classes of light, ordinary, and extra hazard and list under each class typical occupancies in that group. The rule books also set up permissible head spacing and pipe sizes for each group. It is not necessary to memorize all the details, but all fire prevention personnel should gain enough familiarity with them to recognize the different classes of occupancies and sprinkler systems.

B. Change of occupancy.

1. If a more severe hazard occupancy is introduced, the original system must generally be strengthened.

2. Even occupancy changes in the same hazard class may necessitate changes in a sprinkler system. For example, if unusually wide ducts, deep bins, or wide tables are installed, it is usually desirable to add sprinkler heads that will take care of potentially shielded fires. On the other hand, if high piles of stock in a storage area obstruct the sprinklers, storage height should be reduced.

IX. Some basic sprinkler layouts. Figures 12 and 13 show certain basic sprinkler arrangements that should be kept in mind because they will be met with frequently.

Fig. 12. Typical fire department pumper connections: wet pipe system

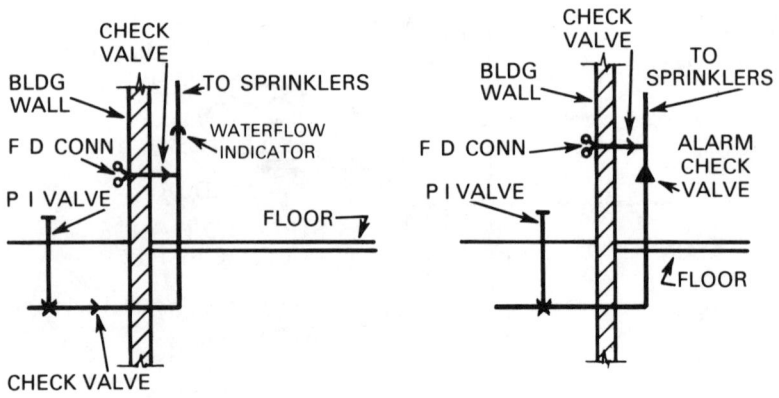

Fig. 13. Typical fire department pumper connections: dry pipe system

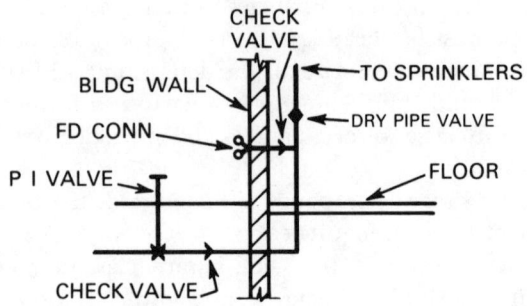

Special Protection Systems

I. Water spray.
  A. Uses. Water spray systems are designed to protect some types of special hazard occupancies and may be automatic or manual in operation. They are not suitable for the protection of buildings and general occupancy. Their principal uses are to protect the types of property listed below.
    1. Oil-filled electrical equipment such as transformers, particularly for outdoor installations.
    2. Storage tanks containing hazardous liquids or gases that may be subject to exposure fires.
    3. Open tanks of certain flammable liquids.
    4. Wall and floor conveyor openings where conditions do not permit the provision of automatic closing doors.
    5. Engine test cells.
    6. Inside generators. Contrary to what most people think, water spray protection inside some types of generators is perfectly feasible and introduces no electric shock hazard. Wet generators can be promptly dried out and restored to service. Such systems do not use nozzles but have special patterns of perforated brass pipe. The systems are manual and are operated only after main and field switches are opened.
    7. Multipass driers. Most multipass driers are protected by sprinkler systems, but, for some types handling large sheets of combustible stock, such protection is extremely difficult to provide except at prohibitive costs. For such installations, water spray protection is a much less expensive acceptable alternative.
  B. Safety considerations. Design of water spray systems is highly specialized because it is necessary to select discharge nozzles of the correct type and then make hydraulic calculations to ensure that the proper discharge is available at each nozzle. Few fire prevention personnel will become involved in this phase of water spray protection. But an inspector should be aware of several basic features that are of vital importance if the system is to function satisfactorily. Among the most important of these items are the following.
    1. Special maintenance.
      a. Spray nozzles have much smaller water passages than sprinkler heads and are therefore much more susceptible to clogging. For this reason, it is usually necessary to have strainers in the water supply lines and to schedule strainer claining as needed.

b. Where foreign materials might enter from outside, nozzles should be equipped with blowoff caps.

c. When water spray piping is painted, nozzles must have protective coverings of a type that will not let any residue through.

2. Inspections and tests. The condition of water spray systems can be pretty well determined by following the inspection and testing procedures described below.

a. When tests are witnessed, be alert to any break in the spray pattern. Breaks may indicate plugged nozzles or piping. Whether or not tests are witnessed, make certain they are conducted annually.

b. Changes in exposure should be checked. (For example, a system designed to protect storage tanks from possible exposure at location A may provide limited or no protection against a new exposure in a different location.)

c. Make a visual examination for such defects as piping corrosion and missing or loose pipe supports.

d. Where a system is operated automatically, make certain the heat-responsive device circuit is in good condition and tested regularly.

e. If the system has operated since the previous inspection, have several spray nozzles removed and examine them for internal obstructions. If foreign material is found, a recommendation for cleaning all spray nozzles is in order.

3. Compatibility. Water spray is compatible with foam, and both may be used together.

4. Drainage. Most installations protecting flammable liquids operations should have provisions for disposing of escaping flammable liquids in a safe location.

5. Hose. At many installations, it is desirable to have a portable hose with water spray nozzles to reach areas that cannot be reached by the stationary nozzles. The hose may be connected to system piping.

II. Foam. Foam fire extinguishing systems have been used for many years to control or extinguish fires in flammable liquids. More recently, their use has been extended to include the extinguishment of some types of fires in ordinary combustibles and the insulation of tanks and important process equipment from exposure. (Foam extinguishing capabilities also are quite frequently important in direct firefighting operations.)

Fire Protection Systems

Foam extinguishes fires by forming a blanket over the burning material, thereby preventing oxygen from reaching the fire. Because foam includes water, it also has some cooling effect. There are various types of foam, and the effectiveness of a foam extinguishing system depends partly on the characteristics of the foam selected, including its degree of adherence, persistence, fluidity, permeability, opacity, and compatibility with other fire extinguishing agents. A system must be designed to use the type of foam that is suited to the hazard involved.

A. Personnel safety.

1. Hazards. Foam systems may be of either the total flooding or the local application type. In total flooding situations, it is possible for personnel to become completely inundated. Even though foam is nontoxic, this type of situation presents three potentially serious personnel hazards.

   a. Visibility may become so reduced that trapped persons cannot find an exit and may stumble blindly into moving machinery, open pits, or other physical hazards.

   b. Floor surfaces of most types become slippery. The possibility of a person slipping and, for example, striking a sharp object cannot be ignored.

   c. There is sufficient air in a foam blanket to permit reasonably adequate breathing through a handkerchief. However, not everyone knows how to breathe under such circumstances.

2. Safety measures. Because of the three kinds of hazards listed above, provision must be made for orderly evacuation of any occupied area where total flooding foam protection is provided. For example—

   a. where there are hazard potentials, such as a machinery pit, that would quickly be covered by the foam, identifying markings should be placed on the ceiling at eye level on walls or columns;

   b. for unusually large areas, prealarms may be needed.

B. Types of foam. As has already been noted, there is a wide variety of firefighting foams. It is not essential to discuss the compositions of these materials in detail. However, fire prevention personnel should be familiar with the basic types of foam. There are two basic methods of generating foam: chemical and mechanical.

1. Chemical foam extinguishing systems. Some foams currently in use are produced by the reaction between ammonium phosphate and sodium bicarbonate in the presence of foaming agents and

water. Foam-making materials should if possible be stored in the temperature range of 50° F. to 90° F. In no case should storage temperatures go below 35° F. or above 100° F. In cold climates, piping for outdoor installations may require insulation and possibly a steam tracer. Steam generation may require a small boiler plant and a heated generator house, which may be so costly that another approach to the problem is warranted. There are three methods for producing chemical foam. Each has its own characteristics and advantages or disadvantages for specific types of operations.

    a. Single powder foam generators. Proportioned amounts of powder are fed into the water stream at a single generator. Foam is produced in the pipe or hose between the generator and the delivery chamber. Foam is applied by water pressure. Some systems are automatic, but most are manually operated. These systems are primarily used to protect large outdoor tanks. The principal advantage is that little manpower is required. The principal disadvantage is that only a limited length of piping may be used. Expanded foam generates a great deal of friction with the piping and therefore cannot be pumped very far.

    b. Double powder foam generation. This device feeds proportioned amounts of two kinds of powder into two separate water streams that are brought together in a mixing chamber where foaming takes place. This mixing chamber is an integral part of the delivery chamber. Most such systems are manually operated, but some are automatic. These systems are normally used for large outdoor installations. System operation requires nearly double the manpower required for a single powder system, but, because foaming does not take place until the delivery point, pipe runs can be much longer than in single powder systems.

    c. Self-contained stored solution foam generation. Two kinds of solutions are stored separately. When the system is operated, the solutions are combined in a pressure container. Foam is produced in the container and in the piping to the nozzles. The foam is propelled by carbon dioxide pressure. Most systems are automatic but have standby manual releases. These systems are used principally to protect inside dip tanks, quench oil tanks, and similar structures. They require little manpower. However, the limited capacity of the storage containers essentially restricts their use to hazards of small or moderate size.

2. Mechanical foam extinguishing systems. Foam may be mechanically generated by mixing a liquid foam concentrate with air and water. Standard mixing rations are 1½, 3, and 6 percent. Systems using these solutions may be designed to produce expansion ratios as low as 6 to 1 or as high as 1000 to 1. (An expansion ratio of 1000 to 1 means that 1 cubic foot of solution forms 1,000 ft. of foam.) The foam concentrate may be stored at temperatures of from 35° F. to 110° F.

In most mechanical foam systems, the air is introduced into the system fairly close to the point of application in order to eliminate foam fluid flow problems that would otherwise result from expansion.

Systems are easily adapted to automatic operation and have the capacity to handle a wide range of flows and to use a wide variety of storage facilities and capacities. These characteristics permit the use of mechanical foam systems for many applications, including fire control and insulation as well as fire extinguishment. (For example, it is possible to use such a system to insulate an important tank against a threatening exposure inside or outside a building while retaining capabilities for fire extinguishment elsewhere.)

C. Uses. Foam systems are most often designed to extinguish fires in flammable liquid. In addition, the systems perform well in protection against exposure and are occasionally used to prevent fires—for example, to spread foam on an airport runway before a disabled plane lands. Some foams with a high expansion ratio and a reasonable degree of persistence have also been used effectively on some types of Class A fires.

1. Foam formulations. Foam generated either chemically or mechanically may be formulated for a variety of uses. The chief types are listed below.

a. Regular foam for hydrocarbon fires. (Such foam breaks down quickly if used on polar liquids.)

b. Special foam for polar liquids—alcohols, esters, ethers, ketones, or any other kind of flammable liquid that will mix with water.

c. Foams especially formulated for use with various dry chemical extinguishing agents.

d. Other special foams required for special conditions.

2. Limitations. Foams are not satisfactory for use on fires involving the following kinds of substances.

a. Combustible metals.

    b. Unusually volatile liquids.
    c. Materials that carry their own oxygen.
    d. Flammable gases.
    e. Liquids heated to 212° F. or higher (for example, some heated quench oil tanks).
  D. Maintenance procedures. Foam systems should be maintained in the same way as other piped auxiliary systems. There are, however, several special requirements, as indicated below.
    1. Strainers in mechanical foam systems should be inspected and cleaned semiannually.
    2. The condition of foam concentrates and foam powders should be tested annually.
    3. Stored solutions for chemical foam systems should be replaced every three years.
    4. Mixing chambers and other mixing devices should be inspected and cleaned after each use.
    5. Cleaning and/or flushing should be carried out as appropriate after each operation.
  E. Special operating considerations.
    1. Foam nozzles and applicators differ in size and discharge characteristics. Substituting one type for another is poor practice.
    2. Automatic systems should include provision for shutting off the water after completion of foam discharge.
    3. High expansion, total flooding systems should have—
      a. high level ventilation to release air displaced by foam;
      b. interlocks to close low level openings and to shut down ventilating systems, flammable liquids pumps, conveyors, and other devices that may spread a fire.
  F. Summation. Foam extinguishing systems play an important role in limiting fire spread when they are properly used and maintained. Inspectors should, therefore, pay particular attention to system maintenance and to any major occupancy change that might require modification of the system.
III. Carbon dioxide.
  A. Characteristics and capabilities.
    1. Carbon dioxide extinguishes fire by displacing part of the atmosphere at or around the fire so that the oxygen content in the area is reduced to 15 percent by volume or less. At this point, there is no longer sufficient oxygen to support combustion in most material. To extinguish fire in a few kinds of materials, the oxygen concentration must be reduced to less than 15 percent—in

some cases, as low as 6 percent. The extent of oxygen depletion required to extinguish fire in such materials is listed in certain handbooks.

2. Carbon dioxide is inert, noncorrosive, nontoxic, and electrically nonconductive.

3. Although carbon dioxide is nontoxic, in total flooding situations it can cause unconsciousness or death through suffocation. Therefore, predischarge alarms are mandatory for occupied areas with total flooding carbon dioxide systems so that evacuation is assured before discharge begins.

4. Carbon dioxide has little cooling effect and is therefore not the best material to use on deep-seated fires or on fires where sufficient residual heat remains to cause reignition once the oxygen concentration again exceeds 15 percent. Carbon dioxide should only be used in such situations if the proper carbon dioxide concentration can be retained in the fire area until sufficient cooling has taken place.

5. At temperatures below $-40°$ F., carbon dioxide becomes sluggish. Nitrogen is added to storage containers to assure effective operation at such temperatures.

6. When carbon dioxide is released from an extinguishing system, it entrains air, which usually contains some water vapor. The low temperature of the discharging carbon dioxide causes this water vapor to condense out. For this reason, carbon dioxide, although an excellent extinguishing agent for most electrical fires, is not recommended for certain types of electronic equipment that are highly sensitive to water damage.

7. Carbon dioxide is useless on fires in materials that carry their own oxygen.

8. Carbon dioxide is not suitable for use on most reactive metals, including sodium, potassium, magnesium, titanium and zirconium. Neither should carbon dioxide be used on the hydrides of such metals.

9. Carbon dioxide is heavier than air. This property is of great value in extinguishing fires in many flammable liquids processes, most electrical equipment, many types of engines, records storage, some food preparation processes, fur vaults, and the like.

10. Some authorities suggest carbon dioxide for use on fires in gaseous flammable processes. However, the majority opinion, to which the author subscribes, is that carbon dioxide should not be used on such fires because it may create an explosive situation.

B. Personnel safety. Fire prevention personnel seldom get involved in carbon dioxide system design. However, those who have life safety responsibility should be alert to certain potential problems in occupied areas with total flooding carbon dioxide systems.

1. Prealarm facilities should include a signal to an attended duty desk, for several reasons. The prealarm bell might not ring, or it might ring too softly to alert occupants. Transients in an area might not be aware of what the alarm is for. Personnel may fall over obstructions and be injured. In addition, accidental tripping of a system while testing or servicing might bypass the prealarm bell.

2. Self-contained breathing apparatus should be available for possible rescue operations. Remember that a 20 percent concentration of carbon dioxide can cause death in 20 minutes, and concentrations in total flooding systems run 30 percent or more.

3. The area involved should be thoroughly ventilated after a fire.

C. Types of systems. Carbon dioxide systems are divided into two basic types: low pressure and high pressure.

1. Low pressure. When carbon dioxide is used at low pressure, an insulated and refrigerated tank is used. Tank pressure is usually about 300 psi and tank temperature 0° F. Low pressure systems are normally used where carbon dioxide is needed for purposes in addition to fire protection or where a fire extinguishing capacity of 1 ton or more is needed.

2. High pressure. When carbon dioxide is used at high pressure, only approved storage cylinders should be used. In using these cylinders, certain general rules should be followed.

    a. Temperature. The permissible temperature range is 32° F. to 120° F. for local application systems and 0° F. to 120° F. for total flooding systems. Artificial heating or cooling should be provided as needed to maintain these temperature ranges.

    b. Pressure release. All cylinders must have frangible discs or other means to relieve excess pressure.

Note. Because pressure increases would result from a power outage that made the refrigeration system inoperative, storage tanks must have relief valves. For the same reason, tanks should have high and low pressure alarms.

D. Installation. Certain installation features are especially important.

1. Location. Containers should be so located that the need for long pipe lines will be minimized. Piping should be protected as required to prevent mechanical injury.

2. Activation. Automatic systems may be activated by any system of approved detection devices, such as heat detectors and smoke detectors. Where a carbon dioxide system and a sprinkler system are both installed, the carbon dioxide system should be set to operate first so as to minimize the potential for water damage.

3. Selector valves. Systems may be designed to protect against more than one hazard. In such systems, fire detectors operate the selector valves to direct carbon dioxide to the proper hazard.

E. Causes of poor performance. Normally, carbon dioxide extinguishing systems are designed by specialists, plans are reviewed by insurance engineers, systems are installed by experienced contractors, and the completed systems are inspected and tested by qualified personnel. Therefore, it can generally be assumed that the original installation was satisfactory. However, records indicate that too high a percentage of these installations either fail to operate in a fire emergency or operate ineffectively. Fire prevention personnel should have an overriding interest in the reasons behind those instances of poor performance. Some of the most common causes are as follows.

1. Construction changes, not noted, that prevent full travel of the manual pull lever.
2. Fuse link actuating devices painted.
3. Line check valves removed for maintenance and reinstalled backwards.
4. Nozzles removed for maintenance and not reinstalled in the same order. Not all nozzles have the same discharge pattern.
5. Actuator tubing so brittle or cracked that pressure cannot be transmitted to the tripping relay.
6. Automatic dampers corroded shut.
7. Nozzle removed for cleaning or other purpose and opening plugged.
8. Extra nozzles added without increasing pipe sizes.
9. Hazard size or type changed without changing system.
10. Building or occupancy changes at ceiling level with no compensating changes in the actuating system.
11. Loose or missing brackets and pipe hangers.
12. Construction of new wall or floor openings for ventilation or work pass-through without means for closing openings during operation of a total flooding system.
13. Changes in building or process ventilation without compensating changes in the system.
14. Selector valves frozen in position.

F. Maintenance. Standard procedures for inspecting, testing, and maintaining carbon dioxide systems are established by the insurance company having jurisdiction or by governmental fire authority. Standard forms are provided for recording this work. Most fire prevention personnel do not, as a rule, have a direct hand in these activities, but they do have the responsibility to see that the required inspection, testing, and maintenance is done and that necessary corrective actions are programmed.

IV. Dry chemical. Dry chemical extinguishing systems are fairly new arrivals on the scene. They serve as auxiliary systems to supplement sprinkler protection, and they may be of either the total flooding or the local application type. Currently, most dry chemical systems use finely ground sodium bicarbonate mixed with other ingredients that assure free flowing and eliminate caking from moisture. However, some systems use other dry chemicals. Carbon dioxide, nitrogen, or other nontoxic gases are used as expellants. There is no completely provable explanation of how the dry chemicals extinguish fires, but the most commonly accepted theory is that the chemicals affect combustion chain reactions.

Dry chemical particles suspended in an expellant gas and flowing through piping do not conform to conventional fluid and gas laws. In addition, particle sizes vary from manufacturer to manufacturer, as do the additives used. For these reasons it is essential that systems be designed by specialists and the approved plans be followed exactly by the installer. No changes should be made in the system or in the dry chemicals used without prior review and approval by qualified personnel.

A. Personnel safety. Discharges of dry chemicals impair visibility and cause trouble in breathing. All dry chemical systems, therefore, must have a predischarge alarm. Where a total flooding system is used and human occupancy is involved, safe evacuation is the primary concern.

B. Uses.

1. Capabilities. Dry chemical extinguishing systems are suitable for fires in the materials listed below.

a. Class B materials. Dry chemicals are especially well adapted to extinguishing fire in hot asphalt or oil because they will not cause boilover.

b. Class A materials that are not subject to deep-seated fires.

c. Electrical equipment other than such highly sensitive equipment as computers and switchboards, in which the powder might insulate fine contacts.

d. Ventilating hoods and ducts for such operations as commercial deep-fat fryers.
2. Limitations. Dry chemical systems are not suitable for use on fires in the following materials.
   a. Class A materials where deep-seated fires might develop.
   b. Materials that carry their own oxygen.
   c. Flammable gases.
   d. Combustible metals. There are dry chemicals that are effective on combustible-metal fires, and some fixed systems have been installed. However, before such systems become widely used, more needs to be done on the mechanics of the various types of delivery systems that will be required.
3. Combination with other systems. Dry chemical systems provide a quick knockdown of fire where they are suitable. However, where deep-seated fires or large bodies of heated flammable liquids are involved, the effects of dry chemical systems must be considered transient, and reliable backup support must be provided. Properly used, dry chemical extinguishing systems are valuable tools in the fire prevention goal of limiting the spread of fire.
   a. Sprinkler systems. In most instances, dry chemical extinguishing systems must be backed up by sprinkler systems. Like a carbon dioxide system, a dry chemical system should be set to operate ahead of the sprinkler system so as to minimize the potential for water damage.
   b. Foam systems. It is frequently desirable to twin dry chemical and foam systems. However, it is necessary that the two types of extinguishing agents be compatible. Some dry chemicals are not compatible with foam, and some are compatible with some kinds of foam and not others. It pays to know the difference.
   c. Other systems. Dry chemical systems may be interlocked with other systems so that, when a fire occurs, alarms sound, air handling systems shut down, industrial processes stop, and other appropriate changes occur. Such arrangements can be quite helpful, provided the interlocks are of the proper type and are maintained.
   d. Maintenance. Basic maintenance procedures are spelled out in the instructions provided by the manufacturer. The inspector should make certain that these procedures are adhered to. Of particular importance are the following:
      1. Checks to make certain nozzles are clear and in proper position, that operating controls are properly set, and that components have not been damaged. (For example, if a pipe

hanger or bracket is missing or broken, the pipe may sag. The change in flow characteristics resulting from the sag could slow the flow of powder and cause clogging in a bend of the pipe.)

2. Annual operating tests.
3. Cleaning of lines after each use or test.
4. Checks of condition of dry chemicals.
5. Proper arrangement of outlets. Like other types of auxiliary systems, a dry chemical system may include a variety of outlets. After these devices are removed for inspection and cleaning, they must be put back in the same order they were taken out.
6. Training program.

V. Halon. Halogen compounds have been around for a long time, but, for all practical purposes, their use for fire-extinguishing purposes dates back only to World War II. Since then, such use has been expanding rapidly. This text will be concerned with the effects of halogens on various types of fire situations and will disregard the rather complicated chemical reactions that take place in halogen fire extinguishment.

A. Characteristics and capabilities. Of all the halogens, only two are normally used in extinguishing systems. They are Halon 1301 and Halon 1211. Both have about the same characteristics. Slightly different concentrations of each are required for fire extinguishment, and Halon 1211 is somewhat more toxic than Halon 1301. Halon 1301 is by far the more commonly used, and so this text will discuss the characteristics of that halogen that are of particular interest to fire prevention personnel.

1. Halon 1301 concentrations of 5 percent by volume or less are capable of extinguishing fires in most surface-burning materials. Such concentrations are very low in toxicity and nonsuffocating.
2. Halon 1301 is colorless and odorless.
3. The chemical is five times heavier than air.
4. Halon 1301 is stored as a liquefied gas under pressure and discharged as a gas.
5. The chemical is a nonconductor.
6. Essentially noncorrosive, Halon 1301 leaves no residue.
7. Pound for pound, the chemical ranks as one of the most effective extinguishing agents available.
8. As normally used, Halon 1301 is not likely to create damaging condensation of water vapor.

B. Toxicity. Halon 1301 concentrations of only 5 percent or less can extinguish fires in approximately two-thirds of the flammable liquids listed in tables available to the author. However, some kinds of fires require higher, more toxic concentrations, and, where the potential for such fires exists, plans for evacuation must be made. Restrictions on human occupancy for various Halon 1301 concentrations are listed in Table 2. Regardless of toxicity ratings, it should be recognized that vapor concentrations near the floor will be higher than elsewhere. For this reason, a prealarm should be provided on all systems.

Table 2

Concentration by Volume	Restrictions on Human Occupancy
0% to 7%	None
7% to 10%	Capability for evacuation in one minute
10% to 15%	Only occasional occupancy and capability to evacuate in one minute.
Over 15%	No occupancy unless self-contained breathing apparatus is available.

Note. It is quite possible that somewhat higher concentrations are safe, but until there is positive proof of this, it is best to be conservative.

As with carbon dioxide, the area involved should be thoroughly ventilated after a fire, and the products formed by halon in reaction with burning materials should be removed.

C. Operation and maintenance. Halon 1301 is used for both total flooding and local application systems. Methods of operating and maintaining the systems are very similar to those described for carbon dioxide systems, and the information will not be repeated here. The differences between halon and carbon dioxide systems are primarily in design and are not of particular concern to most fire prevention personnel.

VI. Summation.

A. Uses of auxiliary fire extinguishing systems. Auxiliary systems are designed to control or extinguish fires in the following materials and occupancies.

1. Flammable liquids and other easily ignitable, fast-burning materials.

2. Ordinary combustibles in high-value or vital occupancies that are unusually susceptible to damage from water, smoke, or heat.

3. Important electrical equipment, especially when downtime is critical.
4. Electronic gear of many types.
5. Special occupancies, such as archives whose contents are peculiarly susceptible to water damage.
6. Engine test rooms or cells.
7. Miscellaneous other sensitive occupancies.

B. Combination with sprinkler systems. Auxiliary systems are generally supplements to sprinkler sysems and are not a substitute for proper sprinkler protection. There are only a few kinds of hazards that can be totally protected by a properly selected auxiliary system—for example, flammable, water-reactive chemicals stored in a separate, fire resistive room or building. Some reasons for the usual supplementary status of auxiliary systems are as follows:

1. Sprinklers can function longer and be restored to service faster than auxiliary systems.
2. Occupancy changes have far less effect on the adequacy of sprinkler protecion than on that of auxiliary systems, and necessary changes can be made more easily.
3. Sprinkler reliability is much superior to that of auxiliary systems. Few sprinkler systems fail to function during a fire. The failure rate for auxiliary systems is much higher, primarily due to substandard maintenance. Few sprinkler systems turn in an ineffective performance during a fire. A somewhat higher rate of auxiliary systems is ineffective under fire conditions. One of the major reasons for this is that the relationship between sprinklers and hazards is better understood. For example, when major occupancy changes are proposed, consideration of possible need for sprinkler changes is almost automatic. Too often, the same is not true of auxiliary systems.

Auxiliary fire extinguishing systems have provided excellent service in limiting the spread of fire. By recognizing their place in the overall picture, and by making certain that high maintenance standards are adhered to and that systems are revised as required by occupancy changes, fire prevention personnel can go a long way towards improving the already good performance.

chapter 7
Occupancy Utility Systems

For the purposes of this chapter, occupancy utility systems are defined as those systems that are installed to meet particular occupancy needs and that normally will not remain with the building if that occupancy is discontinued. Such systems are also known as auxiliary systems. In general, these systems are not under the jurisdiction of a building department. They are far more likely to be checked out by other municipal departments, including the fire service and public safety organizations. In many instances, the systems will also be checked by the technical experts of the insurance company having jurisdiction.

Because building occupancies are subject to periodic changes, it should be anticipated that the utility systems serving the occupancies will also be modified from time to time. It is important that these utility system changes be economically sound as well as technically correct. If an inspector recommends changes that would cost more than the losses they are designed to prevent, or that would impose an undue financial burden, particularly on a small company, the recommendations are not, as a rule, carried out. Fire prevention personnel, although not designers, can provide sound guidance by seeking practical solutions to problems.

In a few instances, special requirements are needed for handling utility system controls in a fire situation so as to limit the spread of fire damage. Operation of the wrong switch or valve may cut off electricity, gas, or air conditioning to occupancies completely outside the fire area. Besides causing an unnecessary loss of production, the loss of power, fuel, or air conditioning can damage some particularly sensitive products and/or processing equipment. Fire prevention personnel should learn to recognize such situations and to make certain that responsible plant personnel have been appropriately trained. Effectiveness in this area does not require any special engineering talent; it does require the application of "horse sense."

The general arrangements of occupancy utility systems are substantially the same as those of building utility systems. For comments on arrangements, the reader is referred to Chapter 5, Building Utility Systems. In the present chapter, attention will be focussed on representative situations that are peculiar to auxiliary systems and in which knowledgeable fire prevention personnel can provide valuable assistance to plants, institutions, and other clients.

Electrical Systems

During the next hour—assuming it to be an average hour—there will be twenty or more electrical fires in the United States. Direct property losses from these fires will be in excess of $60,000. Indirect losses, including lost jobs and lost production, will total $200,000 or more. There will also be several injuries and quite possibly one or more deaths. Most of these fires do not need to happen. The record shows that, in well-run organizations with good fire prevention programs, fires do not occur at anything like the national average. For the most part, the next hour's fires will result from the lack of follow-through on plans to correct defects or from the lack of an economic solution to known deficiencies.

I. Special systems for hazardous locations.
  A. Types of hazards. American industry includes a number of types of operations in which hazardous concentrations of flammable vapors or gases, combustible dusts, conducting dusts, or combustible fibers may be present most of the time, intermittently, rarely, or only when equipment fails or other abnormal operating conditions are present. Some examples of hazardous areas are as follows:
    1. Unventilated pits in rooms where flammable vapors or flammable, heavier-than-air gases are present.
    2. Gas-vaporizing plants.
    3. Areas for the grinding and pulverizing of flour, cocoa, combustible metals, or other combustible materials.
    4. High lint-producing areas of textile plants.

B. Cost reduction. Electrical equipment for such hazardous areas is expensive. Much can be done to reduce costs so that the really essential special equipment becomes economically feasible. Examples of such cost reduction follow.

1. A small room was being used for an operation involving the use of substantial amounts of flammable liquids, and there was a possibility that hazardous concentrations of flammable vapors might form. Recommendations for replacing ordinary electrical equipment and wiring in the room with explosion-proof equipment had not been carried out because the cost was prohibitive for a small plant. Instead, the plant had taken a calculated risk that the ventilation system would not fail and that no accidental large spill would occur. A second inspector took the time to check equipment locations and conditions on the other side of the wall separating the hazard area from other occupancies. He found that virtually all of the electrical equipment and 75 percent of the wiring could easily be relocated to the safe area. The cost of this change was less than 15 percent of the cost of total replacement, and so there was no problem in getting the recommendation carried out.

2. Tool cribs, machine shops, and similar operations are commonly located in lean-to buildings adjoining aircraft hangars. Similar combinations of safe and hazardous areas occur in many other industries. Hazardous areas normally are provided with positive safety exhaust ventilation systems that take suction from within about 6 in. of the floor and appropriate electrical equipment. Adjoining safe areas generally do not have safety ventilation or special hazard electrical equipment. For convenience's sake, it may be decided to make openings in the wall between safe and hazardous areas and to provide fire doors at the openings. A problem is immediately created because, in the event of a major flammable liquids spill, hazardous vapors may be expected to flow into the so-called safe areas. There are two possible ways to solve problems of electrical equipment and wiring in such situations.

   a. Carefully survey the formerly safe area and arrange for replacement or relocation of equipment and wiring as required to meet the hazard. Such alterations may be very expensive, and correction of nonelectrical problems may add to the cost.

   b. Provide a ramped concrete curb of appropriate height at each opening. Properly done, this will prevent the flammable vapors from entering the safe area and eliminate the need for any revisions to the electrical system. The procedure usually is relatively simple and inexpensive.

3. In many cases, lights can be moved outside hazardous areas and illumination provided through transparent panels. In some occupancies, these panels should be wired glass.

4. Motors, in many instances, may be relocated outside the hazard areas. When this is done, motor drive shafts are extended through the walls, and shaft openings are tightly sealed.

5. Switches, controllers, and similar equipment can be installed outside the hazard areas in many cases.

6. If proper attention is paid to wiring arrangements, the bulk of the wiring can be located outside the hazard areas.

C. Types of hazards.

1. Hazard classes. As noted in Chapter 5, the *National Electric Code* sets up three special hazard classes: Class 1 (flammable liquids and gases), Class 2 (combustible and/or conducting dusts), and Class 3 (fibers and lint). Each class has two divisions. Division 1 comprises those areas where the principal hazard problem is present continuously, intermittently, or periodically under normal operating conditions. In Division 2 areas, the principal hazards are not likely to be present under normal operating conditions, but less severe hazards may be expected to be present. Detailed descriptions of the various hazard classes are given in the *National Electric Code* and in such standard texts as the NFPA *Fire Protection Handbook* and the FM *Handbook of Industrial Loss Prevention*.

Electrical equipment designed for special hazard use is readily identifiable by the information on the nameplate. Because this specialized equipment is costly, most purchasers have it installed by qualified professionals. However, occupancy conditions can and do change. When such changes occur, the fire prevention inspector should be particularly alert and keep in mind certain basic principles.

    a. Except in the unlikely event that a piece of equipment has been approved for more than one class, electrical equipment for one type of hazard area should not be used for a different class. For example, Class 2 dust-ignition–proof motors do not belong in a Class 1 occupancy.

    b. Within a hazard class, Division 1 equipment may be used in a Division 2 area, but Division 2 equipment should not be used in a Division 1 area.

Occupancy Utility Systems

2. Modifications and maintenance.

a. Occasionally, a hazardous industrial process will be increased or decreased in size. If an increase in size is substantial, it may be necessary to replace some Division 2 equipment with Division 1 items and to enlarge the Division 2 area. In this sort of situation, it is usually desirable to note the pertinent information and promptly consult with a qualified specialist in the field. Do not guess.

b. Sometimes a process is changed to include the use of hazardous materials different from those formerly used. In such cases, an inspector should determine whether the new and old materials both belong to the same class and subdivision. (Subdivision group is indicated by a letter: A, B, C, and D for Class 1 and E, F, and G for Class 2.) Usually there will be no problem, but in some cases, changes in electric equipment are necessary.

c. From time to time, electrical parts may need to be replaced. To the extent possible, such electrical work should be scheduled for nonoperating periods. If the work is of an emergency nature and must be done during operating hours, plant programming should provide for the elimination of potential sources of sparks or arcs.

3. Dust explosions. Such explosions are infrequent, but a high percentage of those that do occur are disastrous in terms of loss of life, property losses, and/or business interruptions. Major dust explosions may generate pressures that in many instances exceed 100 psi. (Flour, corn starch, and powdered aluminum are capable of initiating these pressures.) Such pressures will knock out walls, raise roofs and floors, smash sprinkler piping, and cause other major damage, and it is highly probable that major fires will follow. By the time the fires are extinguished, major reconstruction will be necessary to restore production.

A sizable number of dust explosions are caused by the use of improper or defective electrical equipment. In checking areas where dusts are present during normal operating periods, inspectors should not limit their attention to the dust-producing equipment. Tops of beams and girders, window ledges, open trays, and other surfaces should be checked for dust accumulations. The reason for this is that many—probably most—of the dust explosion disasters are not a single explosion. Instead, an initial dust cloud is ignited, the pressures generated by this explosion throw a much larger dust cloud into suspension, and more explosions follow. When there are dust accumulations, it pays to be especially careful about the type of electrical equipment around nearby dust-producing equipment.

4. Lint fires. The largest single cause of fires in industries where hazardous amounts of lint are present is electrical equipment failure. The number of electrical fires can be sharply reduced by providing the right type of equipment and then maintaining it properly. A wide variety of equipment is available for Class 3 occupancies. Types of equipment required depend on the amount and type of lint produced, fiber combustibility, and other factors. Some illustrations of factors to be used in selecting equipment are given in the next section of the chapter.

II. Common defects in auxiliary electrical equipment. Installation and maintenance requirements for auxiliary electrical systems are, in general, very similar to those for building electrical systems, and these common requirements will not be repeated here. However, some occupancies have potentially hazardous conditions or electrical equipment that are not common to building systems or to most occupancies. Such hazards include the following.

A. Vibrations. In some occupancies, such as textile mill weaving areas, there is considerable vibration. Where such conditions exist, only flexible cord and conduit that have been approved for the occupancy should be used. (Preferably, motor connections should be flexible cord.) To the extent possible, ordinary flexible connections should be used only in areas with no or limited vibration.

B. Oil. A number of operations require the use of substantial amounts of oil for lubrication. Wiring that is exposed to possible oil contamination should have oil-resistant insulation.

C. Floor leakage. In multistory buildings, it is not uncommon to have wiring attached to the ceiling or located in cable trays that are close to the ceiling. Over a long period of time, occupancies in floors above may have soaked the floors with oils, acids, or other liquids. Exposed insulation on wiring in this type of situation should be resistant to the oil, acid, or other liquids as appropriate.

D. Cable troughs in floors. In some occupancies, cables may be laid in troughs or trenches in concrete floors rather than attached to walls or ceilings. Usually, removable metal plates will be placed over the troughs. In addition to electrical cables, the troughs may contain fuel or water lines, neither of which should be present. Inspectors should also spot-check the troughs for such items as the following.

1. Type and condition of insulation.

Occupancy Utility Systems

    2. Wood blocking. Wood blocks are often used to separate cables when work is being done, and sometimes they are not removed when work is complete.

    3. Combustible debris and/or mice. The author has found both more than once.

E. Mare's nests. Wall, ceiling, and floor outlets are supplied by wiring that is designed to carry a definite amount of current and no more. All too often, far too many lines are plugged into a single outlet. Recently, eleven lines were found tied into a single outlet through a maze of two- and three-way extension outlets. Wiring concealed in the wall was running hot. Had this condition continued for any length of time, there would have been a fire in the wall.

F. Suspended lights. Today, most lights suspended from ceilings are pipe pendant, but there are still some lights suspended on drop cords. Be sure such fixtures have suitable wire guards. Major fires have resulted when an incandescent light was carelessly left in contact with combustible material and not turned off.

G. Ballasts. Many electrical fires have been caused by faulty ballast units on fluorescent lights. Such fires are normally minor if the occupancy is safe and the light is suspended well below the ceiling or mounted on a fire resistive ceiling. However, if the ceiling is combustible (and many are), it should not be possible for the ballast to be in contact with the ceiling.

H. Dirty motors. Inspectors may overlook the fact that open motors that are not located in lint hazard areas can and do pick up sufficient combustible debris to result in fires, unless there is a cleaning schedule appropriate to the occupancy served. Some laundry areas and major sewing machine operations are typical of areas where periodic motor cleaning is essential.

I. Portable electrical equipment.

    1. Life safety. Whenever portable electrical equipment is used around flammable liquids, make certain that the proper equipment is available and that it is not possible for unapproved portable equipment to be present in or available to a hazardous area. The life you save may be that of the next workman you meet. For example: In one small area at one end of a building, plastics were mixed with solvent having a low flash point. Fixed electrical equipment was of the proper type, and approved, portable, explosion-proof equipment had been provided for the mixing operation and for cleaning plastic residues off mixer blades. Unfortunately, this equipment was kept at the opposite end of the

room from the hazardous operation. To save time, a workman used ordinary portable electric equipment for cleaning blades. There was a spark from the motor brushes, and flammable vapors were ignited. The workman became excited and tipped over a half-gallon container of the solvent, much of which spilled on his clothes. The resulting fire was quickly extinguished by the two sprinkler heads that operated. It caused eleven dollars worth of damage to the building and sixteen dollars worth of damage to work clothes—but the workman died three days later from burns.

    2. Extension cords. Extension cords used to supply lights or various types of electrical appliances should have the same current-carrying capacity as the appliance cord. Smaller cords overheat and, in the presence of combustibles, may cause fire.

J. Armored cable. Various types of armored cable are used throughout industry and elsewhere. As long as certain facts are recognized, armored cable provides useful service.

    1. Insulation on some armored cable conductors is combustible (usually varnished cambric). This kind of armored cable should be used only in exposed and relatively dry locations.

    2. In bending armored cable in concealed locations, or in cutting the cable, care must be taken to avoid damaging the conductor insulation.

K. Transformers. Many industrial furnaces employ transformers, some of which may be oil-filled and of considerable size. Units of 75 kva or larger should be in vaults.

L. Circuit breakers. Most circuit breakers used in industry are not subject to regular heavy duty, but certain operations, such as some arc furnaces, subject breakers to extremely heavy use. Unless these units receive above-average maintenance, the plant is flirting with breaker failure and major loss.

M. Telephone switchboards. Major switchboards are highly susceptible to water damage and should be located in a fire resistive or noncombustible room or building. In a combustible, sprinklered building, the switchboard room if at all possible should be made fire resistive or noncombustible and the room sprinklers removed and openings plugged. If the room cannot be satisfactorily altered, the sprinklers should remain in service, and a waterproof canopy should be installed over sensitive portions of the switchboard.

N. Electrically driven fire pumps. These devices must be reliable. More than one plant has been lost because of loss of power to fire pumps at a critical time. Arrangements to assure more complete reliability include the following.

1. If the main switchboard is not exposed to fire, it is permissible to wire the fire pump circuit to the main switchboard, provided the connection is ahead of all secondary circuit fuses or circuit breakers.

2. If there is any likelihood that the main switchboard may become involved in a fire, the fire pump circuit should be connected directly to transformer secondary leads and ahead of all disconnecting means.

O. Refrigerators. A number of severe refrigerator explosions have resulted from the storage of flammable liquids and gases in ordinary refrigerators. Only explosion-proof refrigerators should be used for such purposes. Many medical and chemical research facilities must refrigerate small amounts of flammable liquids and gases, mostly in glass bottles and test tubes.

P. Raised floors. Many computer rooms and similar areas have fairly large raised floors, under which is generally a maze of wiring. This wiring may or may not have combustible insulation. The raised floor is usually metal, but it may be wood or have wood supports. In addition, most raised floors have openings near some pieces of equipment, and so combustible debris can get under the floor. For all of these reasons, representative panels should be temporarily removed during each visit to permit an inspection of what is below the floor.

III. Summation. Throughout modern America there are tens of thousands of electrical installations designed to do many kinds of work. By and large, they perform their functions efficiently and play a major role in our standard of living. Few of the electrical fires that do occur result from any complicated sequence of events. Most of the fires stem from inattention to such fundamentals as misuse of equipment, essential inspections, and maintenance, some examples of which have been given above. Fire prevention personnel can do much to improve the loss record by developing habits of careful observation that will generate the ability to quickly recognize potential problems.

Air-Handling Systems

Most air-handling systems are essentially building systems, but certain kinds of occupancies typically have separate systems. Failure or misuse of one of these auxiliary systems can lead to serious property losses and, in some instances, to loss of life. Fire records for auxiliary air-handling systems have been generally good, but there is room for improvement based on a better understanding of what can go wrong.

I. Process cooling. A number of occupancies, some of them critically important, require an essentially constant temperature and humidity and/or freedom from dusts. Some also require the movement of large volumes of cooled air to remove the heat generated by the operating equipment. Of course, it is generally desirable to keep the temperature and humidity fairly constant and to free the air from excessive dusts, but variations of 5 percent or more do not seriously affect most operations. However, auxiliary systems that are designed to serve occupancies with specific requirements must maintain tighter control of system operation and should therefore be independent of building systems. There is no generally accepted term for such specialized systems, but this chapter refers to them as process cooling.

  A. Types of systems.

    1. Unit systems. The term unit system as used here refers to a system serving one room or one occupancy area and not to window air conditioners. Unit system air conditioning machinery may be located in the area served or in an adjoining air conditioning machine room. The advantage of this type of system is that it has no ducts that pass through floors, walls, or other areas that may be susceptible to fire. The lack of such ductwork minimizes the risk of damage from fire, smoke, or heat from sources outside the room.

    2. Independent systems other than unit systems. These systems have ductwork that passes through other areas, and the air conditioning machinery may be located away from the area served. The extent of special precautions necessary depends on the degree of exposure from the occupancy and/or building structure through which the ductwork passes. There should be no openings in the ductwork between the machine room and the special occupancy areas. In addition it may be necessary to take the following precautions.

      a. Install an automatically closing fire damper in the wall between the special occupancy and the exposing occupancy.

      b. Wrap the exposed ductwork in a covering having the desired fire resistance rating.

  B. Important components.

    1. Outside-air intakes. Locations for outside-air intakes are normally selected so as to minimize the possibility of drawing

combustibles or other foreign material into the system. When such special operations as sensitive electronic equipment or fine bearings are involved, it is essential that intake location receive particular attention. In some instances, it may be necessary for heat detectors to shut down the air conditioning system, the operating equipment, or both in the event of an exposure fire.

2. Return-air intakes. It is costly for an air conditioning system to use 100 percent outside air, and so recirculated air forms a substantial part of the air demand in ordinary systems. If the return-air duct from a special occupancy area must pass through another occupancy, the duct may require protective measures similar to those required for supply ducts.

3. Air filters. Preferably, air filters should be noncombustible and of a type that removes particles down to a designated size. If the filters are of the proper type, the primary concern of the fire prevention inspector is the adequacy of the maintenance program. Throwaway filters must be replaced at regular intervals and other types regularly cleaned. Failure to adhere strictly to schedule could cause an unacceptable amount of combustibles to reach sensitive equipment.

4. Thermal cutouts. Many operations involving electronic gear, particularly in the computer field, develop dangerously high temperatures in the absence of a continuous flow of cool air. The effect could be a serious fire or heat distortion of critical equipment. In either case, there would undoubtedly be loss of computer memories, and this loss could well affect operations far removed from the damage area. To prevent this from happening, many units have a thermal cutout. These devices, in addition to other possible capabilities, are designed to deenergize electronic equipment whenever the air supply fails.

5. Air conditioning shutoffs. The air conditioning for some types of computer systems must continue to run for a time after the electronic equipment is deenergized. If it does not, excessive temperatures build up. For such systems, manual air conditioning controls should be located where they are immediately available to properly trained personnel.

In some cases, temperatures as low as 150° F. cause malfunctioning of some components. Temperatures below 500° F. can so impair system reliability that major replacements become necessary. When the potentials from smoke damage are also considered, it become obvious that not much of a fire is required to do heavy damage.

6. Building system ductwork. If possible, none of the ductwork of the building air conditioning system should pass through a computer room or another occupancy with a special system. Any building system ductwork that does pass through the special occupancy area should be encased in material having a fire resistance equal to that of the room's walls, floor, or ceiling.

C. General safety considerations. It is probably safe to say that most special occupancy air conditioning systems are properly installed, because the equipment they protect is quite expensive and its loss would cause far-reaching problems. The air-handling systems generally are carefully designed, the plans are reviewed in detail by experts, and the installation is accomplished under qualified supervision. It is what comes afterward that creates most of the problems for fire prevention engineers. For example an inspector must be concerned with the following questions.

1. How good is the maintenance system?
2. Are operating personnel well-trained in emergency procedures?
3. Have there been any building or occupancy changes that affect the integrity of the system?

II. Exhaust safety ventilation. The general run of air-handling systems is designed to provide conditioned air or a desired number of air changes per hour for a building or a specific area. These systems are equipped to discharge part of the circulated air outside the building. In many instances, particularly in the larger installations, controls are provided to permit rerouting of air flows, so that the products of combustion may be discharged outside the building in case of fire.

Not included in the general air-handling class is another type of system whose functions under normal operating conditions are to safely remove hazardous materials from the buildings, to reclaim usable materials, and to serve other purposes that are not within the scope of this text. Many of these systems are relatively small, but this does not mean that they do not have the potential for serious damage if factors such as those listed below are neglected.

A. Dusts.

1. Explosion prevention. As has been noted earlier, combustible dusts, depending on the degree of fineness, can when ignited develop dangerous explosion pressures. To minimize this risk potential, ignition sources must be prevented from entering the system. For example, smoking should be controlled, magnetic separators should be installed near system inlets where ferrous tramp metal

might be present, and mesh screens should protect inlets where foreign materials might get drawn into the system. Depending upon the layout and the type of equipment served by the system, there may also be a need to control the possible development of friction sparks.

2. System layout. Even if an installation has a satisfactory layout to begin with, it may not stay that way. It must not be lost sight of that equipment is frequently relocated, and/or machinery may be added or removed. Such changes require alterations to the exhaust system, and the plant may make these modifications without a review of plans by qualified personnel. Items to be considered include, but are not limited to the following.

   a. Materials used. These should be of the same quality as the original installation.

   b. Bends. Sharp bends should be avoided and the number of bends kept to the minimum possible.

   c. Inspections. There should always be enough handholes to permit inspection and cleaning of a pneumatic pipe or duct.

   d. Collectors. Any new collector should, if possible, be located on the roof or elsewhere outside the building.

   e. Fans. New fans should be nonferrous, and any needed fan housing should be either nonferrous or lined with nonferrous material. If possible, fans should be located on the clean air side of the collectors.

   f. Pipe or duct ridges. To the maximum extent possible, interior surfaces of pipes or ducts should be free of ridges that would collect appreciable amounts of dust.

   g. Explosion venting. All new pipes or ducts should have explosion vents appropriate to the material being handled.

3. Toxic materials. Some dusts are toxic in addition to being combustible. When such dusts are present, under no conditions should the air from the "clean air" side of the collectors be recirculated.

4. Collectors.

   a. Outdoor dust collectors handling combustible dusts, if important to plant production or a potential source for major fire damage, should be protected by automatic sprinklers. In cold areas, the sprinkler system should be dry pipe or nonfreeze as appropriate.

   b. In general, indoor collectors handling combustible dusts should be protected by automatic sprinklers.

c. Where magnesium dusts are present, water wash collectors should be used. The installation should be appropriately vented to dissipate the hydrogen generated by the wetting of magnesium sludge.

5. Grinding. In metal grinding, certain precautions are necessary to avoid problems in the exhaust systems. For example:

a. Machines used for grinding magnesium or magnesium alloys should not be used for grinding other kinds of metal.

b. Before machines used for grinding, buffing, or in otherwise processing one type of material are used for another material, they must be cleaned thoroughly.

B. Combustible particles other than dusts. Exhaust systems are used to remove combustible wastes other than dusts. Quite common is the removal of shavings and chips from woodworking operations. Fire, rather than explosions and/or fire, is the basic danger here. It is essential that tramp metal be removed before it enters the system, smoking be controlled, and regular checks be made to prevent clogging of the system.

C. Noncompatible materials. Two or more ventilating systems are sometimes siamesed together in a common exhaust system. There is no objection to this provided the materials exhausted are compatible. In chemical laboratories and in some commercial operations, noncompatibles may meet in the common exhaust. Fire or explosion is then almost inevitable. An example of this hazard is the combination of a strong oxidizer, such as perchloric acid, with organic materials.

D. Escalators. Originally, escalators were enclosed in shafts similar to those used for stairs. Because of merchandising demands, most escalators in stores are now open. Many means have been developed to protect escalators. Of these, one in particular involves air-handling systems. Basically, this system operates by creating a downdraft through the opening of sufficient volume to prevent smoke and hot gases from rising from floor to floor. Simultaneously with the downdraft an exhaust ventilation system may be put into play to remove smoke and hot gases from a fire floor, discharging them outside the building. The systems may be triggered by a pilot sprinkler line, heat-actuated devices, or smoke detectors. These systems are not too common, but they are in use.

E. Flammable liquids and gases. The forms of safety exhaust ventilation systems most often seen by the average inspector are probably those used to remove hazardous amounts of flammable-liquid vapors or flammable gases. These systems are provided for a wide range of operations, including paint-spraying and dipping, engine testing, and multiple burner gas-fired ovens. Many of these systems must tie in with other safety equipment in order to function properly. Such tie-ins are covered in chapters 8 and 9, which relate to specific types of operations and equipment. In addition, there are some basic considerations that are unfortunately often overlooked. Among the most important of these are the following.

1. Vapor density. The vapors of all flammable liquids are heavier than air, and so are some flammable gases. Other flammable gases are lighter than air. Operating processes change at times, and it is entirely possible for a lighter-than-air gas to be introduced into an area that has an exhaust ventilation system designed for the removal of heavier-than-air vapors. For the system to be effective under the changed conditions, exhaust duct intakes must be relocated so that they are above the process giving off lighter-than-air vapors. It may also be necessary to relocate fresh-air inlets for the area.

2. Multibranch systems. It is not uncommon, particularly in some forms of engine testing, for duct branches leading from several test stands or several operations to come together in a single exhaust duct leading to the outdoors. When heavier-than-air-vapors are being exhausted and the systems are poorly designed, it is entirely possible for hazardous vapors to flow back through ducts to nonoperating engines or processes. The author has seen several instances in which it was necessary to install positive means in the ducts to prevent such backflow.

3. Compatibility. Not all gases or vapors are compatible. Where incompatibles are present and operations require ventilation, separate systems should be provided.

4. Spills and ruptures. Safety exhaust ventilation systems are not designed to take care of vapors from large-scale spills, important ruptures, or runaway chemical reactions. Where these things can happen, the proper procedure is to look for ways to eliminate or at least minimize the possibility. The ventilation system will help, but it can't do it all.

5. Air intakes. Occasionally it is necessary to take fresh air for one area from other rooms within the building. Where this is necessary, make certain that the intake is not near a place where flammable vapors are present or where there are gas burners. (At one plant, when all the spray booths in a room were operating, fresh

air was admitted by leaving the door to the next room open. Close to the door was a moderate-sized oven heated by ribbon gas burners. Repeatedly, when the spray booth fans were turned on full, the gas flame was extinguished, and raw gas escaped. Pending the construction of a new intake, a windbreak or shield was provided to divert the air draft.)

6. Positive or natural ventilation. An operation that produces hazardous amounts of vapor generally needs positive mechanical ventilation if the flammable liquid has a closed-cup flash point of 110° F. or less. Positive ventilation is also needed if the liquid has a higher flash point but is heated to or over that temperature. Under other circumstances, natural ventilation is generally adequate. However, if topography and prevailing winds sometimes create a backflow through the exhaust system, positive ventilation is essential.

It should be noted that temperature limitations given above are based on average ambient temperature conditions in the United States. Where ambient temperatures are higher, consideration should be given to raising the temperature limits.

F. Summation. Safety ventilation systems that are properly designed, installed, and maintained and that are modified as required to meet changes in operating procedures play a vital role in minimizing the possibility of fire and in limiting the extent of fire should one occur.

chapter 8
Industrial Furnaces, Ovens, Driers, and Incinerators

A large percentage—perhaps a majority—of products in current use were subjected to high temperatures at some point in the production process in order to obtain the desired results. The hot processing includes such diverse operations as heat treating, paint drying, cloth drying, bright annealing, coffee roasting, lumber kilns, fiberboard drying, and bake ovens. Temperatures used range from a few hundred to several thousand degrees. Evaporated fluids range from low–flash point flammable liquids to water. Work-chamber atmospheres may consist of ordinary air, flammable gases, or inert gases. Equipment may consist of small, more or less standardized, units, or it may be several hundred feet long. Work may be stationary or travel through the equipment on a conveyor. Most processes use oil or gas for fuel.

Because of the wide variety of temperatures, materials, and equipment involved in hot processing, it is necessary to define terms. For the purposes of this chapter, furnaces are defined as those units that operate at temperatures in excess of 700° F. in the work chamber. Ovens and driers are defined as those units that operate at temperatures of 700° F. or lower in the work chamber.

Safeguards described are those that are basic. For large and/or complicated units, it is customary for the insurance carrier to work with the company involved to set up a procedure for inspecting and testing each unit. The carrier and company decide, among other matters, which controls must be tested and how often, the sequence in which tests should be made, and how tests are to be made and evaluated.

This information, together with an overall report on general conditions, should be at each furnace, oven, or drier. Inspections and tests should be made by qualified personnel only. In addition, standard operating procedures should be posted at each important unit and adequate provision made for their enforcement.

At many locations, some materials must be disposed of by burning. For this purpose, well-designed incinerators are necessary. Incinerators may be small units or large enough to occupy a fair-sized building.

Process Furnaces

Process furnaces are typically used for vitreous enameling, annealing, heat treating, and similar processes. Fires and/or explosions in such furnaces originate chiefly in the fuel system and/or in flammable gas in the work space.

I. Fuel systems.
   A. General.
      1. Operators. Furnace operators should be both thoroughly trained and competent in the following procedures.
         a. The correct operating procedures in starting up, holding on the line, and shutting down.
         b. Actions to be taken in emergency situations.
         c. Inspection, testing, and maintenance procedures.
      2. Safety controls. All safety controls and their application should have approval from a recognized testing authority such as UL or FM. An inspector should make sure that the controls are maintained, that none is bypassed, and that none is taken out of its original designed sequence of operation. This last check is not as complicated as it sounds. Trained operators can explain the sequence and the reasons for it.
      3. Trial for ignition. Normally, a time limit (usually about five seconds) is prescribed for lighting off. If the flame does not become established in that time, the lighting-off procedure must be started over. Most furnaces must be purged before each successive start in order to prevent a hazardous fuel-air mix from accumulating.
      4. Pilots. Before opening any burner cocks or valves, operators should make certain a reliable pilot (automatic or manual) is in front of each burner outlet.
      5. Fail safe. In the event of failure or malfunction of a vital component, an automatic or semiautomatic system should automatically shut down in the proper sequence.

B. Gas systems. Almost all recorded gas system explosions occurred during lighting off or firing.

1. Lighting off. Nearly two-thirds of recorded explosions in gas systems happened during initial or secondary lighting-off operations. Of these, almost all occurred when operators failed to close individual burner cocks and provide stable pilot flames prior to opening main furnace gas valves. There is much to be said for systems arranged so that burner cocks must be closed before a main gas valve can be opened.

Most multiburner systems are required to have a device providing a preventilation purge period before lighting off. Such devices should never be bypassed. A preventilation setup has a time delay relay that holds back other operations for a predetermined period. The relay setting should never be changed by operating personnel.

2. Firing. Slightly more than a third of recorded explosions in gas systems occurred during firing. To the best of the writer's knowledge, none of these explosions occurred when work-chamber temperatures were over 1,400° F. Temperature dials should be looked at during inspections. Principal causes of trouble during firing periods are as follows:

a. Flame failure with no means for fuel safety shutoff. Combustion safeguards will prevent this situation.

b. Inadequate combustion air. To prevent this situation, interlock the combustion air and the fuel supply so that the system will shut down in the event of excessive fuel pressure or the failure or serious partial failure of the air of fuel supply.

c. Furnaces operating at under 1,400° F. should have continuous pilots. For other furnaces, the warming-up period is the key time when special care should be taken.

C. Oil systems. There have been fewer explosions in oil-fired systems than in gas-fired systems, not because oil systems have less hazard potential, but because there are fewer of them. Principal causes of oil-system explosions have been the following:

1. Flame failure with no means provided for automatically shutting off the fuel supply.

2. Lack of interlocks between fuel supply and combustion air. Such interlocks shut down the system if either supply fails.

3. Oil leakage through partly open valves into a warm work chamber without a pilot. This is similar in effect to the unsafe practice known as "firing off the walls" which is sometimes used to light heating or power boilers.

    4. Failure to properly purge a unit before making a second trial for ignition.

II. Special-atmosphere furnaces. Some heat-treating processes require special atmospheres in the work chambers. In some instances, the special atmosphere is an inert gas, in which case there is no particular problem. In other cases, the special atmosphere is flammable. For such units, appropriate provision must be made to eliminate the explosion hazard in the equipment used to provide the special atmosphere. Most furnaces of this type are individually designed, and the operators are provided with detailed instructions. Such operators should not only be well trained in correct operating procedures, but also be qualified to inspect, test, maintain, and repair equipment.

Wherever feasible, furnaces with special flammable atmospheres should be completely purged with inert gas before starting operations and after stopping the process. If this cannot be done, it is necessary to burn out the air during startup operations and to burn out the flammable atmosphere when shutting down. Individual units differ in the exact methods of accomplishing these two essentials, but the following basics normally apply.

  A. Furnaces that operate at 1,400° F. or higher.

    1. Interlock the supply of flammable atmosphere with the temperature controls so that the atmosphere cannot enter the furnace until operating temperatures are reached.

    2. Keep operating temperatures above 1,400° F. while burning out the special atmosphere.

  B. Furnaces that operate below 1,400° F.

    1. After operating temperatures are reached, locate reliable pilot at outlets for flammable gas.

    2. Start gas flow. When gas ignites, pilot may be removed.

    3. When atmosphere burns at both exit and entrance points, provide continuous pilots at these points.

    4. Burning-out operations differ somewhat and should be conducted in accordance with design instructions.

III. Summation. The criteria in preceding paragraphs provide a general guideline which should be supplemented as needed to fit safety requirements for individual units. Specialists in the field should refer to the detailed publications in this area. For the generalists, it is usually enough to know—

  A. the general principles involved;

  B. whether or not operators are well trained and are following instructions;

Industrial Furnaces, Ovens, Driers, and Incinerators 145

C. whether or not there is an adequate inspection, test, and maintenance program.

Ovens and Driers

Ovens and driers are used in many types of industry and range in size from a small, box-type unit to equipment several hundred feet long. Material being processed may be stationary or carried on a conveyor, which in turn may travel slowly or quite rapidly. These variations must be taken into account by the designer arranging controls to assure safe operation. This section will outline certain basic criteria that are applicable to all or most ovens and driers, to specific methods of operation, or to selected classes of equipment.

I. Siting. Virtually all ovens have inherent danger possibilities—for example, ventilation failure, fuel escape, overheating, and overload of combustibles. A fire or explosion resulting from such occurrences can shut down a plant or injure personnel in addition to causing direct property losses. For these reasons, ovens and driers should be so located that the risk will be minimized.

Ovens and driers, particularly large units, should, if possible, be located in one-story noncombustible or fire resistive buildings or in the first story of a multistory building. Because of the problems of emergency access and ventilation, basements and other below-ground locations should be avoided. If ovens or driers must be located above the first story, waterproof the floor as necessary to prevent leakage into the story below. Other siting precautions include but are not limited to, the following.

A. Separate ovens and dryers from nearby hazardous operations, preferably by means of fire cutoffs.

B. Keep combustible materials and/or operations involving combustibles a reasonable distance from the oven or drier.

C. Make sure clearances are sufficient to permit adequate discharge from sprinkler heads, if present, and for the use of hoses and/or extinguishers.

II. Construction. In order to minimize potential hazards resulting from construction details, the following features should be included in all new ovens and driers and, where possible, incorporated into existing units.

A. Construction materials should be noncombustible.

B. Interior finish should be smooth to minimize the collection of combustibles and for easy cleaning.

C. Oven roofs and floors should be insulated, and spaces above and below should be ventilated so that their temperatures remain within acceptable limits.

D. If an oven or dryer must rest on a combustible floor, a concrete slab of suitable thickness or its equivalent should be installed under the oven or drier.

E. Panel joints on the work area side of roofs and walls must be vapor tight so that condensable vapors cannot get inside the panels.

F. Ducts from ovens should be insulated as necessary and have adequate clearance when passing through combustible roofs, floors, walls, or ceilings. Tables for such requirements may be found in the recognized building codes.

G. At a number of types of operations, there are vapor explosion hazards. Where such hazards exist, explosion venting is needed. This can take the form of venting panels or, on small batch ovens, the provision of explosion venting or of friction latches on outward-swinging oven doors.

III. Ventilation. Any time the operating temperature of an oven exceeds the flash point of the flammable vapor being driven off, there is a potential explosion hazard. To eliminate or minimize this hazard, safety ventilation is provided. As a rough rule of thumb, sufficient ventilation should be provided to keep the vapor concentration at or below 25 percent of the lower explosive limits, except as noted below. See Chapter 9. Basic guidelines for safety ventilation are as follows:

A. Batch ovens. This type of unit drives off a large part of the vapors in the first few minutes of operation. During that period, the vapor concentration approaches the lower explosive limit. To reduce the concentration by 75 percent for this brief period would be prohibitively costly. Instead, most authorities require the following.

1. Ovens operating at 250° F. or less should have ventilation that can handle 380 cfm referred to 70° F. for each gallon of volatiles introduced in the maximum work load (this refers to maximum volatiles load).

2. Ovens operating at over 250° F. need ventilation to handle 530 cfm referred to 70° F. per each gallon of volatiles introduced.

The above ventilation rates may be modified when exact calculations are made for the oven temperature actually used.

Usually, safety ventilation is correctly provided for the originally programmed work load. However, conditions change, and inspectors should be alert to increases in the volatiles work load or in oven temperatures. Such changes may require alterations in ventilation.

B. Continuous ovens. Not less than 10,000 cu. ft. of air referred to 70° F. should be provided for each gallon of volatiles entering the oven.

IV. Electrical equipment. Many ovens and driers have electrical equipment inside the unit. This equipment and other equipment in the immediate vicinity should be suitable for hazardous locations as defined in the *National Electric Code.*

V. Some common oven and drier fire causes. There have been fires in virtually every type of oven and drier known to man, the frequency, of course, varying with the degree to which each type is used and the extent of the hazard. A careful observer could have predicted a high percentage of these fires in advance, because rarely were they the result of anything unusual or complex. Some of the most common causes of oven and drier fires and explosions are listed below.

A. Spontaneous heating (perhaps this should also be called poor housekeeping). Deposits of such combustibles as paint, lacquer, and lint are allowed to build up for too long. In many instances, such deposits heat up to the point where ignition takes place. Do not make the mistake of believing that spontaneous heating takes place only in flammable liquids. For example, cotton lint packed down beside a steam radiator gradually carburizes. As it does, the temperature required for ignition goes down.

B. Careless handling of combustibles. Many fires result when combustibles accidentally come in contact with burner flames.

C. Sparks. Defective electrical equipment is the chief source of fire-causing sparks. Static and friction sparks play a minor role.

D. Inadequate or poorly maintained safety controls. At times, a safety device shuts down an oven repeatedly, indicating that something is seriously wrong. Operators, particularly those on piece work, have been known to bypass or shunt out such a device or to block it from operating. (A folded piece of cardboard in the right place will block most mercoid switches.) Operators do this because they lose money while the unit is down for repair. Some people apparently don't mind risking their necks for a few bucks. During inspections, it pays to take a look at representative controls. More than once, the author has found several devices shunted out at a single operation.

E. Fuel system failure. Causes for fuel system failure in ovens and driers are similar to those described for furnaces and will not be repeated here.

F. Overheating. This information could be included in item D above, but fires from overheating occur frequently enough to warrant separate mention. Various defects and accidents can cause equipment to overheat so much that fire ensues. Some of the most common are listed below.
   1. Deteriorated thermocouple.
   2. Control set too high.
   3. Control mechanism defective.
   4. Steam pressure too high. Inspectors should be especially alert to this in textile and fiberboard plants where loose combustible debris can lodge on or behind steam coils.
   5. Accidental breakage of infrared lamps.
   6. Replacement of parts with substandard equipment or failure to replace at all.

VI. Fire protection. Automatic sprinklers should be provided for most large ovens or driers that process appreciable combustibles, include appreciable combustible construction, or are likely to accumulate combustible wastes. The sprinkler system should have a readily accessible control valve. In most instances, the system should also protect the exhaust ducts. Automatic deluge or automatic water spray systems should be provided for ovens or driers that operate at higher temperatures than rated sprinkler heads can protect and for certain other types of driers. Special protection systems backed by sprinklers are sometimes used. Whatever protection method is used, the inspector should satisfy himself or herself that the system is in service and is being properly maintained. A sprinkler head with a heavy deposit of paint on it is not of much use.

VII. The operator. Unquestionably the most decisive factor in whether or not an oven or drier is safe is the operator or occasionally, as will be seen, the watchman. No matter how carefully an operation is designed, no matter how many safety features are built in, the program may fail, and fire, perhaps of gigantic proportions, may ensue unless the responsible employees are both well trained and reliable.

   A. Case 1. Fall River, Massachusetts, 1941. A former textile mill had been taken over by a major rubber company for the manufacture of various articles essential to the war effort. In one of the upper stories of a main building was a large, sprinklered vulcanizer. The night watchman on his rounds heard the sprinkler alarm and, on investigating, found two sprinkler heads had operated and were controlling a small fire in the vulcanizer or drier. He waited until he thought the fire was out, closed the sprinkler control valve to prevent further water damage, and went on about his rounds without doing anything further.

On his next round, the watchman found that flames had broken out again and now formed a large fire. At this point, he got excited and called the fire department, but he forgot about the closed valve. Although the fire department responded promptly and plant officials arrived shortly thereafter, it was some time before anyone realized the control valve was shut. By this time, it was too late for the sprinklers to do any good, because far too many heads had operated for the available water supply. There were other contributing factors, but the watchman's error was the principal reason a fire that should have cost $5,000 or less resulted in a multimillion dollar loss. It might also be pointed out that, with a war on, a substantial portion of all the crude rubber in the United States was destroyed. Had synthetic rubber not been developed shortly before, this loss might easily have prolonged the war and thus inevitably increased war casualties and property losses.

B. Case 2. A lithographing oven in New Jersey was equipped with all necessary fire safety devices, and these were well maintained. On the upstream side of the oven was a well arranged printing setup.

At quitting time, one of the operators started to clean the rollers at the printing presses before he shut down the oven. The solvent used for cleaning the rollers was naphtha, which was contained in an ordinary can.

Investigation after the loss clearly indicated that, in a moment of carelessness when his attention was diverted to other matters, the operator accidentally placed the naphtha can on the conveyor. Since it was still moving, the conveyor carried the container into the still-operating oven. Heat from the oven vaporized the naphtha far more rapidly than the oven safety ventilation system could evacuate it. Since ignition sources were present, there was an explosion that destroyed the oven and decreased production for an extended time.

Had the oven heating system been shut down first, or had the conveyor been stopped prior to cleaning the rollers, this explosion would not have happened.

VIII. Summation. Fire prevention personnel other than specialists cannot be expected to become expert in all the varied types of ovens and driers—or, for that matter, process furnaces. Personnel do need to know or obtain the following information.

A. The principal causes for fires and explosions in ovens, driers, and furnaces.

B. Whether or not the equipment has been designed and installed by qualified personnel.

C. Whether or not the work load or process has been changed since the original approval.

D. Whether or not inspections, tests, and maintenance are being conducted in accordance with the requirements of the authorities having jurisdiction.

E. Whether or not operators have been properly trained.

F. Sources of information for problems that are outside of the inspector's training and experience.

Incinerators

One of the problems of the industrial age is the disposal of rubbish. Part of the trash is burned, and devices to accomplish this range from small, back yard trash cans to large units capable of handling substantial amounts of refuse. It is with these latter units that fire protection personnel are primarily concerned.

I. Basic requirements. Incinerators in industrial, institutional, and large residential establishments should meet the following standards.

A. The incinerator room should be cut off from the rest of the building by fire resistive walls, floors, and roofs. Openings should be protected by automatic or self-closing fire doors of appropriate rating.

B. Ducts and chimneys should be installed and maintained in accordance with the local building code. The *National Building Code* or its equivalent provides good guide-lines for establishments not covered by a code.

C. Trash collection rooms for incinerators should be cut off from other building areas and sprinklered.

D. Refuse chutes should terminate in trash rooms and never in combustion chambers. Openings between trash rooms and refuse chutes should have hopper doors to prevent flashbacks up the chute. (Not too many years ago, a woman shook an empty flour bag into a chute, presumably planning to use the bag for something else. Because the chute led to the combustion chamber and the lower chute door was not of the hopper type, fire flashed up the chute and ignited the woman's clothing. She died of burns.)

E. Refuse chutes should have sprinkler protection at the top and be periodically checked against clogging. (Not long ago, fire broke out in a clogged rubbish chute that had questionable doors. The fire spread up the chute out of reach of the sprinkler at the top of the chute, went through the chute door, and entered a hospital corridor. Sixteen patients died and in addition, there was substantial damage to a technically fire resistive hospital.)

II. Summation. Incinerators are responsible for a great many fires, substantial property losses, and some fire deaths and injuries—but they don't have to be. An exact breakdown of incinerator fires is not available, but it is safe to say that about 90 percent of all the important incinerator fires occurred because at least one of the standards cited above was not met. An inspector can help eliminate fires of this type as important factors in the overall fire loss record simply by acquiring some basic principles and applying common sense.

chapter 9
Flammable Liquids and Gases

Most people, including those in the fire prevention field, have more contact with flammable liquids and gases than with any other type of hazardous materials. In studying the use of flammable liquids and gases, it is well to bear in mind the following simple facts. No material in itself is dangerous. It is the use to which a material is put and how it is handled that determine whether or not that material is hazardous. Properly used, flammable liquids and gases provide us with light, heat, power, and the ability to produce a wide range of useful products not otherwise obtainable. Improperly used, these same liquids and gases can be responsible for major disasters. For example, an automobile engine that is well designed and maintained and properly used is an example of the controlled use of gasoline. If the same liquid is used for cleaning in the open in an unprotected, combustible building with combustible occupancy and with ignition sources present, that use is improper. Under such circumstances, fire is inevitable. The only question is when.

Flammable liquids and gases are rarely responsible for initiating fires, but, in the presence of an otherwise harmless spark or other ignition source, they form an important contributing factor to many major fires. Misuse of flammable liquids alone is responsible for 15 percent to 18 percent of the industrial fire losses. Misuse of flammable gases accounts for a smaller percentage of industrial losses, but it is still a major factor in the loss picture. The two categories are responsible for a disproportionate share of the lives lost in fires. To make a substantial reduction in the losses now being incurred, it is of vital importance that fire prevention personnel have a basic grounding in the how and why of flammable liquid and flammable gas fires.

Flammable Liquids

I. Properties. If flammable liquids are to be controlled so that they serve mankind efficiently and with a minimum of hazard, it is necessary that certain of their properties be understood.

A. Flash point. A flammable liquid's flash point is the lowest temperature at which it gives off enough vapor to form a flammable mixture with air near the surface of the liquid or within a test container. There are two kinds of tests for flash point: closed cup and open cup. A substance usually has a slightly higher flash point in an open cup than in a closed cup test. To be on the safe side, the general practice is to use the results of closed cup tests. It should also be noted that the rate of vapor production goes up with the temperature. Requirements for handling and ventilation thus depend on both the flash point and ambient temperature, as illustrated below.

1. A liquid with a flash point of —50° F., such as gasoline, requires special care, particularly with respect to ventilation, at any normal working temperature.

2. A liquid with a flash point of 90° F., such as amyl alcohol-n, poses no great problem where the temperature does not exceed 70° F. There are intermittent, short-term problems where temperatures occasionally exceed 90° F., and special care is required where ambient temperatures are frequently higher than 90° F.

3. A liquid with a flash point of 150° F., such as fuel oil no. 6, causes special problems only in those few areas where temperatures of 150° F. or higher are part of the operation.

B. Fire point. This is the lowest temperature at which a flammable liquid in an open container gives off enough vapors to continue to burn once it has been ignited. Most substances have a fire point slightly higher than their open cup flash point. Fire points are primarily of interest to industrial chemists and specialists. It is, however, important that fire prevention personnel understand the difence between fire point and flash point and not use fire point data when the actions to be taken depend on the flash point.

C. Explosive range. This term includes all concentrations of flammable-vapor-and-air or flammable-gas-and-air mixtures in which a flash will occur or through which a flame will travel if the mixture is ignited. When the percentage by volume is below the lower explosive limit, it can be said that the mixture is too lean to burn. When the percentage is above the upper explosive limit, the mixture is too rich to burn. The range is usually expressed in percentages by volume. Percentages vary widely, depending on the substance in question, as shown in Table 3. The limits listed are for average conditions. Some variations occur when atmospheric pressures differ appreciably from normal.

Table 3   **Explosive Range Table**

	Lower Limit %	Upper Limit %
Carbon Disulfide	1	50
Divinyl Ether	1.9	36.5
Gasoline	1.3	7.6
JP4	0.8	6.2
Methyl Alcohol	5.5	36.5
Toluene	1.3	7

If vapor-air mixtures within the explosive range are confined and there is an ignition source, there will be an explosion. Explosive pressures are greatest at the center of the range and taper off at each end. For example, a 25 percent carbon disulfide-air mix could generate destructive explosive pressures, whereas a 1 percent or a 50 percent carbon disulfide-air mix would not. The effects of range are sometimes illustrated by a diagram similar to that shown in Figure 14. The explosive range of a material is a key factor in determining the amount of safety ventilation needed and the extent to which venting of explosive pressures may be necessary.

Fig. 14. Explosive range

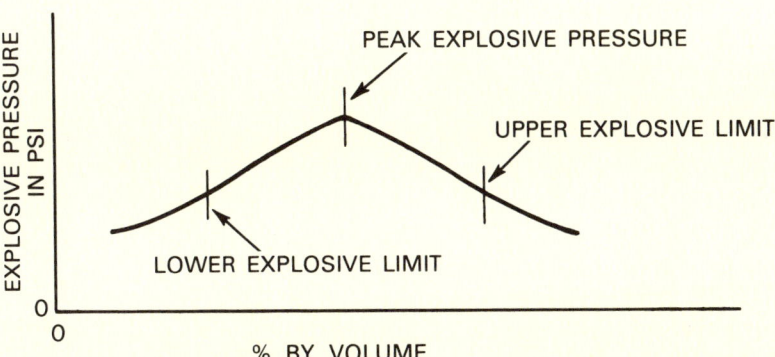

D. *Autoignition temperature.* This term refers to the lowest temperature to which a substance must be heated to initiate self-sustained combustion in the absence of a spark or flame. This critical temperature is influenced to some extent by the nature, size, and shape of the container and by the rate of heating, but most tables list autoignition temperatures for average conditions. Therefore, it is wise to be a little conservative in estimating hazards from autoignition. In most instances, appropriate safeguards for other hazard properties also take care of the autoignition problems, but, at times, the autoignition temperature is the critical factor. For example, a number of industrially used flammable liquids have autoignition points well below 300° F. (Carbon disulfide's autoignition temperature is 257° F.) Such liquids may be used in areas where there are metal surface temperatures well over 250° F.

E. Spontaneous heating. A number of flammable liquids combine readily with the oxygen in the air, which process generates heat. Whenever the heat is produced faster than it can be dissipated, the temperature rises and ignition eventually occurs. Oil or paint-soaked rags in closed or nearly closed containers annually provide numerous instances of this fact of life.

Oxidation rates range from slow to very high, depending on the material involved. Such rates provide the key to the proper requirements for storing and disposing of waste materials—for example, requirements regarding storage containers, methods of storage, and frequency of waste disposal.

There are a number of published tables that show oxidation rates (or susceptibility to spontaneous heating) of various materials. The tables also provide information relative to storage containers and precautions that should be taken to avoid spontaneous heating. One such table which also lists solids susceptible to spontaneous heating is published in the NFPA *Fire Protection Handbook*.

F. Specific gravity. The weight of any liquid compared to an equal volume of water as "1" is its specific gravity. Liquids heavier than water have a specific gravity greater than 1, and liquids lighter than water have a specfic gravity of less than 1. Most but not all flammable liquids are lighter than water and float on the water's surface. A few are heavier than water and sink below the water's surface. This is vital information and affects requirements for fire extinguishing agents, storage, overflow drains, and other safety precautions. For example, fires in carbon disulfide (sp. gr. 1.263) may be extinguished with carbon dioxide, dry chemicals, or water. Fires in gasoline (sp. gr. 0.7) may be extinguished with carbon dioxide, dry chemicals, or foam. Foam is useless on carbon disulfide because the flammable vapors break through the foam blanket and reignite. It should also be noted that, because of reignition and other problems, a carbon disulfide fire requires twice the amount of carbon dioxide or dry chemicals as a fire in the same volume of gasoline.

Trapped overflow drains leading to a safe location are needed in many installations using lighter-than-water flammable liquids. On the other hand, overflow drains are not necessary for a container that holds a liquid that is heavier than water, assuming proper application of water for extinguishing purposes. In that case, the only spillage would be water.

Specific gravity has a number of effects on storage requirements. For example, it is entirely feasible to store carbon disulfide in an open container provided a layer of water of appropriate thickness is placed over the CS_2.

G. Vapor density. This term measures the relative densities of vapors and gases compared to air as "1." The vapors of all flammable liquids are heavier than air, whereas many combustible gases are lighter than air. This is why ventilation for flammable liquids should be at or near the floor level. For duct exhaust systems, inlets should be from 4 to 6 in. above the floor. It is also why ventilation for flammable liquids tanks should be taken from the edge of the tank. Where there is a high rate of vapor formation, it may also be necessary to provide floor level ventilation at the tank.

Ventilation for duct exhaust systems may be natural draft, but, where vapor production is rapid or there is a possibility of a backdraft, positive mechanical safety ventilation is needed.

Ventilation for flammable gases that are heavier than air is similar to that for flammable vapors. Gases that are lighter than air need ventilation at the ceiling level.

Fresh air intakes should be located so that they will assist the safety ventilation process.

H. Vapor pressure. This is the pressure exerted by the vapor in a vapor-air mixture in a closed container when a condition of equilibrium has been reached. This equilibrium is the point where the amount of vapor leaving the liquid is equal to the amount of vapor returning to the liquid. In general, the lower the vapor pressure, the easier it is to get into the explosive range. Probably the outstanding examples of this are certain of the jet propulsion (JP) fuels, which have extremely low vapor pressures and are nearly always in the explosive range. This fact has required a radical redesign of aircraft fuel tanks and revision of standards for fuel pumping. All pumping generates some static, but, with most fuels, the sparks are dissipated harmlessly through vapor-air mixes that are too rich to burn. With

JP4, however, it is necessary to regulate pumping rates and pipe sizes to prevent development of static discharge sparks that can ignite the JP4-air mix. While difficult of proof, it is quite likely that some unexplained jet aircraft fires and explosions that took place prior to the redesign of fuel systems resulted from lightning strikes that would not have affected a "too rich to burn" gasoline-air mix in a similar system.

I. Boiling point. This temperature is of interest primarily to those engaged in refining and distilling operations. However, fire prevention personnel use boiling points to define subdivisions of flammable liquids. If other critical items are taken care of, there will be no fire problem from the boiling point alone.

J. Water solubility. A number of flammable liquids, including alcohols, ethers, and ketones, are completely or partially soluble or miscible in water. Mixing these liquids with water reduces their flammability and eliminates static hazards. Some flammable liquids, such as alcohols, become completely nonflammable when sufficiently diluted.

II. Classifying flammable liquids. Several methods are used to classify flammable liquids. Any of the methods used by nationally recognized authorities is acceptable, but the most commonly used method is probably that shown in Table 4.

Table 4

Class IA	Flash point below 73° F. Boiling point below 100° F.
Class IB	Flash point below 73° F. Boiling point at or above 100° F.
Class IC	Flash point at or above 73° F. but below 100° F.
Class II	Flash point at or above 100° F. but below 140° F.

Note. Boiling point affects vapor production.

The reader will note that the table lists only liquids with flash points below 140° F. This is because liquids with higher flash points are normally classed and treated as combustible liquids and not as flammable liquids. Combustible liquids are the so-called safety solvents.

A number of solids (tallow, for example) have relatively low melting points and, as used in industry, are often found in the liquid state. In such cases, these solids should be treated as combustible liquids.

III. Basic safeguards. Basic safeguards for flammable liquids are listed below. These safeguards provide basic fire safety if used properly with reference to the flammable liquids classification and the amount of flammable liquids involved.

A. Isolate the hazard. For example, a fire in a 55-gallon drum of

Flammable Liquids and Gases

flammable liquid located outdoors on a sandpile would result only in the loss of the liquid. If the drum were in a small, suitably detached shed, the loss would at most comprise the shed and its contents. However, if the drum were in a main building, the fire might involve other contents and the building itself. In such instances, the extent of potential damage is limited only by the extent to which the building and contents are combustible and by the presence or absence of proper protection facilities of adequate capacity.

Suitable locations for bulk storage of flammable liquids include, but are not limited to, the following.

1. Buried tanks.
2. Outside, above-ground tank or drum storage well away from important buildings. Storage areas should be diked and/or provided with safe drainage as necessary. (Not too long ago, the author found a detached open storage shed for flammable liquid drums on raised ground near a road. In event of fire, burning flammable liquids would have entered a storm sewer line from which they would have emerged about half a mile away in a main railroad freight yard. A disastrous railroad car fire might have resulted. The plant involved has corrected this situation.)
3. Detached sheds or small buildings that do not expose important buildings or yard storage. Preferably, these buildings should be of fire resistive or noncombustible construction.
4. Properly cut off rooms, located if possible in the first story. If storage must be on an upper story, provide for drainage to a safe location and make floors liquid tight.

B. Confine the liquid. Procedures and equipment should be arranged to keep flammable liquids inside equipment, piping, or storage containers. The escape of materials should be kept to a minimum and provision should be made for the safe removal of material that does escape. Some devices for confining flammable liquids are listed below.

1. Safety cans.
2. Dip tanks with automatic-closing covers and trapped overflow drains leading to a safe location.
3. Ramped curbs between flammable liquids rooms and other areas.
4. Floors pitched to trapped floor drains leading to a safe location.
5. Diversionary or enclosure dikes for outside storage, as appropriate.
6. Proper storage containers. Glass jars and bottles should never

be used except for small amounts of flammable liquids, such as laboratory samples. Elsewhere where absolute chemical purity is a must, glasslined standard storage containers may be used.

C. Ventilate. Ventilation is essential to the prevention of flammable liquids fires and of possibly serious vapor-air explosions. Ventilation systems, no matter how simple or complex, are designed to confine, dilute, or remove the maximum normal amount of vapor release. They are not designed to completely remove a hazardous vapor-air mix resulting from a major spill or accidental discharge, although they will help. Where such incidents are likely to occur, additional safeguards are required.

In general, it is desirable to design the ventilation system to keep the vapor concentration at or below 25 percent of the lower explosive limit. To accomplish this, it is usually necessary to provide positive mechanical ventilation for flammable liquids with a flash point of 110° F. or less. Positive ventilation is also required for flammable liquids with flash points up to 300° F. if these liquids are heated almost to or above their flash points. For other flammable liquids, natural draft is usually sufficient.

D. Provide explosion venting. Low flash point flammable liquids require explosion venting when used in certain processes or in situations where major spills can be anticipated. Explosion venting can take a number of forms, including spring or friction latches on doors, skylights, wall panels, and other structural items; appropriate combinations of pressure-relieving and pressure-resistant walls; paper diaphragms; and rupture discs. Experience and technical discussions can help an inspector learn where explosion venting is needed and whether or not various forms of venting in use are serviceable.

E. Eliminate ignition sources. Wherever flammable liquids are used, all ignition sources not necessary to the process should be removed to the maximum extent possible. Potential ignition sources that cannot be removed and are not necessary to the process must be controlled so that the hazard is eliminated. Some potential ignition sources are listed below.

1. Open flames, such as cutting and welding torches, furnaces, driers, heaters, and so on. These sources should be kept away from flammable liquids operations. For example, cutting and welding on flammable liquids equipment should not be permitted until the equipment has been properly emptied and purged.

2. Electrical equipment such as d.c. motors, ordinary switches, circuit breakers, and so on. Such equipment should be moved out of the hazard area or replaced with suitable explosion-proof equipment.
3. Overheating—for example, in quenching oils or varnish ovens. Well-maintained temperature controls are in order for such locations.
4. Mechanical sparks.
5. Spontaneous heating.
6. Static sparks.
7. Friction sparks.
8. Smoking. This can and should be prohibited.

F. Provide fire protection. The first five basics cited will greatly reduce the incidence of fires and their severity. However, even these controls may sometimes fail, and potentially major fire situations may result. Installed fire protection is necessary to guard against such a contingency. Wet pipe automatic sprinkler protection is the basic fire protection system for flammable liquids. Depending on the nature of the flammables and how they are used, such auxiliary systems as foam, carbon dioxide, or dry chemical may also be provided.

G. Provide an inert atmosphere. Occasionally it is necessary to keep an inert gas over the surface of a flammable liquid. For example, this safeguard, called *inerting,* may be provided in flammable liquids tanks that for one reason or another cannot be fully emptied and/or purged.

H. Educate and train. All personnel handling flammable liquids should be given a good grounding in the hazards of the materials they are dealing with and in the proper actions to take in emergencies.

IV. Storage facilities. The two principal forms of flammable liquids storage are drums and tanks. Either type may be located inside buildings or in outside storage locations, subject to definite limits on permissible amounts by class. Within buildings the amount permitted will vary according to the story used. For appropriate limits the fire prevention inspector should refer to the limits set forth in the code used and to any additional limits imposed by the insurance carrier having jurisdiction. Aside from limitations on amount, other basic considerations for storage facilities include the following.

A. Indoor drum storage.

1. Storage locations. Locations for indoor drum storage are listed below in order of preference.
    a. Detached, fire resistive or noncombustible building not exposing any important building or equipment.
    b. Attached, cut off building or room.
    c. Ground story of multistory location. Avoid basements.
2. Explosion venting. Where Class I flammable liquids are used and an explosive vapor-air mixture may form in a relatively confined area, means for venting explosion pressures are needed.
3. Drainage. Means must be provided for promptly removing spilled or burning flammable liquids. Trapped floor drains leading to a safe location are the most usual form of drainage. Wall scuppers are acceptable if burning flammable liquids would not expose anything else.
4. Dispensing. If dispensing is done, drums should be grounded. Drums used for dispensing should be equipped with approved drum pumps or self-closing faucets.

B. Outdoor drum storage. Outdoor storage locations must be kept free of weeds. Storage areas should not slope towards buildings, equipment, or sewer and drainage systems. Such areas should be subdivided by wide fire lanes for fire department access. Hydrants with adequate water supplies and adjacent hose houses should be strategically located.

Note: Never forget that an unpurged empty drum or a partially filled drum is far more dangerous than a full one.

C. Storage tanks. Building codes in most areas set up well-defined limits for the location and spacing and size of both buried and above-ground tanks, depending on the class of flammable liquids stored. In outlying areas not covered by building codes, comparable restrictions are imposed for insurance purposes. An inspector should of course make sure tanks meet code requirements and a number of other considerations of importance should also be checked out regularly, for example:
1. What is actually in the tank? It is not unknown for a plant to use tanks for flammable liquids other than those for which they were originally intended. The substitution may be all right, but occasionally it is not.
2. How good is the maintenance? (For example, is protection provided for gage glasses, vents, and similar items?)
3. Are the drainage systems operable?

Flammable Liquids and Gases 163

    4. Is the electrical equipment suitable for the flammable liquid now in use?
    5. If a foam protection system is present, is it operable?
    6. Is the inerting system, if present, operable?
    7. Are diked areas free of weeds?
    8. Are standard operating procedures posted and enforced? (This requirement is particularly important for operations pumping flammable liquids in or out of storage tanks.)
    9. Are floating roofs functioning so there is control of potential vapor spaces?
    10. Are tanks purged before reconditioning?

V. Pumping and piping systems.
  A. General safety measures. Many industrial processes involve moving flammable liquids automatically from one area to another inside the same building or from one building to another. Various safety measures are necessary for such systems.
    1. A requirement that employees be trained and educated in operations and in emergency procedures.
    2. The provision of proper fire protection.
    3. The operation of a continuing program of inspections and maintenance by qualified personnel.
    4. The elimination of ignition sources.
    5. The provision of means for prompt shutdown of operations in the event of a leak.
    6. Provision of other safeguards as appropriate. For example, it is desirable to color code pipes according to the liquid they carry and to post conspicuous signs identifying emergency controls.
  B. Transfer methods.
    1. The safest method of moving flammable liquids is by positive displacement pumps operating under a lift and located away from important areas. Other kinds of pumps may be used if proper precautions are taken (e.g. positive shut-off).
    2. Gravity transfer is acceptable provided automatic safety shut-off valves are located as close to the source as possible.
    3. Hydraulic transfer is suitable for flammable liquids that are not miscible with water.
    4. Inert gas transfer is the least desirable method, but it may be used under strict supervision.
    5. Whatever transfer method is used, exposed piping should be protected against mechanical injury. Piping should never run

through troughs or chases containing service lines or through pits.

VI. Mixing operations. Among the most common mixing operations are those found in industries manufacturing paints, lacquers, varnishes, or adhesives. Mixing operations are also carried out on a smaller scale in such processes as the compounding of some pharmaceutical products. Both fire and explosion hazards may be present, depending upon the flammable liquid involved, the method of handling, the degree of confinement, the provisions for ventilation, and the temperatures of the area and the liquid.

A. Hazardous conditions. Mixing operations present all the usual fire hazards of flammable liquids. Also, room explosion hazard potentials may exist at ambient temperatures under these conditions.

1. The flammable liquids being used have flash points of 20° to 25° F. or lower.

2. Flammable liquids that have flash points of up to 100° F. or even slightly higher are or may be heated appreciably above their flash point.

3. A single piece of equipment occupies an important part of the room or area and presents an equipment explosion hazard.

B. Safety measures. Mixing operations should have very careful preplanning, including consultations with qualified specialists in that field. Particular attention should be paid to the following factors.

1. Siting of buildings or rooms.

2. Construction of buildings, including drainage and explosion venting as necessary.

3. Venting of equipment that presents an explosion hazard. Where adequate venting is not possible, equipment should be inerted.

4. Ventilation.

5. Elimination of ignition sources.

6. Heating systems. Systems must not only introduce no ignition sources, but they also must operate at temperatures below the autoignition temperatures of the liquids.

7. Fire protection.

8. Equipment for monitoring temperatures, pressures, liquid levels, and other conditions. Preferably, this equipment should not include sight glasses.

9. Electrical equipment.

C. Summation. There are far too many variables in mixing operations to attempt to deal with them all in detail in part of one chapter. However, an inspector can successfully help safeguard such operations if he realizes that a mixing operation by its very nature produces far more flammable vapors than most other flammable liquids operations. The primary problem is the safe handling of these vapors. In many instances, this will require the provision of safeguards beyond those required for most types of flammable liquids operations.

VII. Dip tanks. Dip tanks, properly installed, normally create no serious problems. Unfortunately, there have been too many instances of neglect of fundamental precautions such as the following.

A. Dip tanks should be installed away from pipe trenches or other low spots where vapors can collect.

B. Tanks should have overflow drains equipped with sealed traps using mineral oil or water. For large dip tanks, emergency bottom drains with quick-opening valves should also be provided. All drains should be piped to a safe location. Drainage provisions should conform with local codes. As a rough rule of thumb, dip tanks over 10 sq. ft, in area should have overflow drains, and tanks of over 500-gallon capacity should have emergency bottom drains.

C. Proper freeboard should be maintained.

D. Pumps and/or agitators should be of the proper type and provided with *accessible* emergency shutoff controls.

E. Automatically closing covers must be operable. A wrongly positioned cover is useless. So is one that has a fusible-link tieback in the wrong position or that could be obstructed by work in the tank.

F. Fixed fire protection systems must be in operating condition.

G. Drainboards must be installed so that, in event of sprinkler operation, excess water will not run into the tank. Figure 15 shows two kinds of drainboard setups for dip tanks.

Fig. 15. Satisfactory dip tank and drainboard arrangements

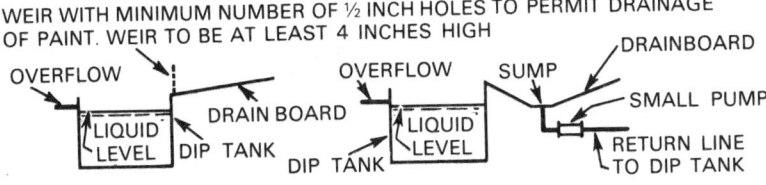

H. Salvage facilities must be provided for unusually large tanks. Some dip tanks hold 20,000 gallons or more. Loss of the entire contents would be extremely costly, and such losses can be avoided with proper salvage facilities.

VIII. Paint spraying. There are many types of paint-spraying operations, and each has its own designed safety features. However, virtually all types have the following requirements in common.

A. Never locate a paint-spraying operation so that there is any possibility of an operator being trapped by a fire. The author has seen several paint-spraying booths in which the operator had his back to a blank wall. If a flash fire should occur in such a booth, the operator might never get out. At best, he would sustain severe burns.

B. Maintain a regular cleaning schedule for booths and ducts. Usually, a thorough cleaning of booths after every forty hours of operation is satisfactory. Ducts should be checked regularly through cleanout openings and cleaned as necessary.

C. Provide for the prompt removal of paint-soaked rags and waste. Use covered metal containers for temporary storage.

D. Protect sprinkler heads from paint loading by the use of grease or light paper bags. Such head coverings should be renewed or replaced when booths are cleaned.

E. Use nonsparking tools.

F. Make certain ventilation system is on before paint spraying is started. A good way to do this is to interlock spray guns with ventilating fans.

G. Make certain electrical equipment fits the hazard.

H. Make sure that paint-spraying operations of any size are suitably cut off from other occupancies that are combustible or are susceptible to damage from smoke, heat, or water.

I. Provide watertight floors and adequate drainage.

J. If possible, install ramped curbs at room entrances.

K. Make certain standard operating procedures prohibit switching from one material to another without first cleaning the booth.

L. Prohibit paint spraying in the open under most circumstances.

IX. Paint driers. Most paint drying is undoubtedly done in either batch or conveyor ovens such as those discussed in Chapter 8. However, a third type of drying oven is in fairly common use, particularly where automotive vehicles are being painted. For want of a better name, this type of oven or drier is often called a tunnel oven. The usual arrangement is to have the paint spray booth and the drier in tandem, with appropriate separation between. A vehicle is brought into the spray room, painted, allowed to stand for a predetermined time, and then moved into the drier for finishing. Heat in a tunnel drier is most commonly generated by banks of infrared lamps. Insulated-sheath resistance heaters with reflectors are also used and are equally acceptable. Infrared driers should be protected by automatic sprinklers when their contents would require such protection in any

Flammable Liquids and Gases

other type of drier. Tunnel driers present two kinds of hazards in addition to those usually associated with flammable liquids.

  A. Preferably, vehicles should be towed into and out of both the spray booth and the drier to eliminate potential ignition sources. Where this is not practical, exceptional care must be taken to see that motors are not started when there is any possibility that a dangerous concentration of flammable vapors is present.

  B. Infrared lamps, properly installed, are a satisfactory source of heat for many drying operations, but they may furnish an ignition source if broken. These lamps must be protected as necessary against mechanical injury.

X. Summation. There are far too many ways in which flammable liquids are used to permit a detailed review of each. For example, flammable liquids in quantity are found in refineries, vegetable oil extraction, quenching oil operations, fabric coating, paint manufacture, pulp and paper industries, and many other operations. The approaches to each hazard must be tailored to fit the situation, but a great deal of information is available to the fire prevention officer charged with such responsibility.

  A. The basic safeguards previously detailed apply to all flammable liquids operations.

  B. There are many sources of information on the characteristics of various hazardous materials.

  C. Codes detail the types of electrical equipment required.

  D. Various sources rate the relative effectiveness of different types of fire-extinguishing agents for each flammable liquid.

  E. Realistic inspection and test requirements can be established.

  F. Various references give information on the hazards to look for.

  G. Limit quantities of flammable liquids. In general, flammable liquids in an operating area should be kept to an absolute minimum (usually one day's supply). Flammable liquids in the plant supply pipeline should also be kept to a minimum. Backup storage on the premises should be limited to the amount needed to assure continuity of operations.

Flammable Gases

I. Properties. In many respects, flammable gases present types of fire hazards that are very similar to those generated by flammable liquids and their vapors. Flammable gases and flammable vapors also have many properties in common. There are, however, some marked differences that warrant special attention—for example:

  A. Flammable gases do not have flash points or fire points when used as gases. These gases at very low temperatures become liquids and at that point acquire flash and fire points. Such liquefied flammable gases are rarely used except in laboratories and the cryogenics field, and so it is unlikely that the flash points of these gases will serve any useful purpose for most fire prevention personnel.

B. Most flammable gases have an explosive range far wider than those of flammable liquid vapors. This fact can and often does influence ventilation design and the extent to which special electrical equipment must be provided. The explosive range may also greatly affect the amount and type of explosion venting required.

C. Whereas all flammable liquids vapors are heavier than air, only some of the flammable gases are heavier than air. Vapor densities for flammable gases range from 0.069 for hydrogen to 2.15 for vinyl chloride and even higher for some other gases. When lighter-than-air gases are used, requirements for the location of ventilation and explosion-proof electrical equipment will differ sharply from those necessary for flammable liquids vapors and heavier-than-air flammable gases. When an inspector deals with flammable gases, he must know the vapor density. Do not guess—it could be a serious mistake.

D. All flammable gases are lighter than water.

E. Many flammable gases are soluble in water to some extent.

F. Spontaneous heating is not a problem with flammable gases.

G. Flammable gases are highly compressible. Flammable liquids are only slightly compressible. To illustrate this basic difference and its potential effect on a fire situation, it is necessary to reemphasize the fundamental gas laws and their applications.

  1. The basic gas laws are as follows:

    a. The volume occupied by a gas varies inversely with the absolute pressure, if the temperature is constant. (Absolute pressure is gauge pressure plus 14.7 psi).

    b. The volume occupied by a gas is directly proportional to the absolute temperature, if the pressure is constant. (Absolute zero is $-459°$ F.)

  2. From the above laws are derived three formulas.

    a. $P_1 V_1 = P_2 V_2$.
    b. $T_1/T_2 = P_1 V_1 / P_2 V_2$.
    c. $V_1 T_1 = V_2 T_2$.

  3. Two examples of the applications of these laws are provided below.

    a. A 100-cu. ft. container is used for the storage of flammable liquid X at 1,000 psi. The container breaks, relieving the pres-

Flammable Liquids and Gases

sure. Since liquids are only slightly compressible, the flammable liquid now occupies 101 cu. ft. or less.

b. The same size container holds flammable gas Y at 1,000 psi. This container also breaks, relieving the pressure. The gas expands to a volume of 6,900 cu. ft.±. $(1,000 + 14.7) \times 100 = 14.7 \times V_2$. Since the gas will mix readily with the air, it is obvious that a much greater area will be exposed to an explosive mix unless provision is made for safely dissipating the gas. In calculating the dangers of flammable gas, the room volume is important. In a small room, it is conceivable that, even if pressure is declining, enough might remain to take out a wall. The same would of course not be true of a released liquid.

H. Representative examples of flammable gases and their explosive ranges, flash points (where available), vapor densities, and water

Table 5

	Lower Explosive Limit (% by vol. in Air)	Upper Explosive Limit (% by vol. in Air)	Flash Point	Vapor Density (Air = 1.0)	Water Solubility
Acetylene	2.5	81.0		0.9	VS
Butane-n	1.9	8.5	−76° F.	2.0	S
Dimethyl Ether	3.4	26.7	−42° F.	1.62	VS
Ethylene	2.7	34.0	−185° F.	0.97	SS
Formaldehyde	7.0	73.0		1.07	S
Hydrogen	4.1	74.2		0.069	SS
Hydrogen Sulfide	4.3	45.5		1.18	VS
Methane	5.3	13.9		0.55	SS
Methyl Chloride	7.6	19.0		1.78	VS
Propane	2.2	9.6	−156° F.	1.56	S
Vinyl Chloride	4.0	22.0		2.15	SS

VS = very soluble   S = soluble   SS = slightly soluble

solubilities are provided in Table 5. Although flash points are listed for some gases, it should be noted that at normal ambient temperatures and in low to moderate pressure containers they are all gases.

II. Some common industrial gases. A wide variety of compressed and liquefied gases is in use throughout industry, in institutions, in living quarters, and elsewhere. A substantial number of these gases are flammable and some are also toxic, corrosive, reactive, and/or unstable. The flammable gases whose primary hazard is not combustion are dealt with in Chapter 10.

Detailed lists of flammable gases and their properties are published by the National Fire Protection Association, The Factory Mutual System, and other nationally recognized authorities. Many of these gases have rather limited usage, but others are widely used in many areas. Table 6 shows the most commonly used flammable gases and some of their major areas of use.

Table 6 — Common Industrial Flammable Gases

Gas	Some Principal Areas of Use
Acetylene	Cutting and welding, chemical mfg., paint mfg., plastic mfg.
Ammonia	Refrigeration, fertilizer mfg., generation of hydrogen, nitric acid mfg.
Gas, light, fuel	City gas and manufactured gas ⎫ light, fuel
City gas (Manufactured)	City gas and manufactured gas ⎭
Ethyl chloride	Anesthetic, solvent
Ethyl ether	Anesthetic
Ethylene	Anesthetic, fruit ripening, refrigeration, synthetic rubber
Ethylene oxide	Fumigation, sterilization
Formaldehyde	Dyes, disinfectants
Hydrogen	Hardening fats and oils, petroleum cracking, wood alcohol mfg., heat treating, special blowtorches

As the table shows, flammable gases are used in many kinds of operations. Quantities may vary from an ounce or two in a household aerosol can to containers rated in thousands of gallons of water capacity.

III. Liquefied petroleum gas (LPG). One of the commonest forms of flammable gas in use today is LPG. LPG actually comprises several kinds of gas, including propane, butane, and pentane. All are heavier than air and lighter than water, and all have high vapor pressure. The gases are, for all practical purposes, colorless and odorless. Domestic supplies have a stench added to aid in detecting leaks, but it is not usual to add this to industrial supplies.

A. An LPG tank should conform to the requirements set forth in NFPA standards 58 and 59 or in publications of the authority having jurisdiction. These standards cover tank construction, sizes, spacing, location in respect to buildings, and other matters. Some of the more common problem areas are described below.

1. Heat from an exposure fire can cause a storage tank to rupture if it has an unwet crown. A ruptured tank would immediately release a large vapor cloud that will ignite almost instantly. To prevent such happenings, there must be exposure control. The exposure should be either eliminated or reduced to the point where it is no longer a hazard. If the exposure cannot be removed or satisfactorily reduced, the LPG tank should be protected by an

Flammable Liquids and Gases 171

automatic water spray system or by some other, equally effective, method of tank protection.

2. For any one of a number of reasons, tank relief valves from time to time discharge vapors which, if ignited, form a torch. Heat from such torches can severely expose buildings and other property. Some appropriate safety measures are listed below.

   a. Make sure there is adequate separation between tank and buildings, yard storage, and other tanks.

   b. Control ignition sources.

   c. Never permit a tank relief discharge to include a U bend. A few years ago, the relief valve on an LPG tank in a Texas town torched. The torch, which was pointing downward, weakened tank seams, and the tank ruptured. The resulting fireball killed nine firemen and injured a number of other people.

3. Changes made since the original installation may have introduced hazards.

   a. Make certain ground level still pitches away from important buildings, yard storage, or other vulnerable areas.

   b. Make sure that both containment and diversion dikes are maintained and that the dikes and the spaces between them and the tanks are free of unnecessary vegetation.

   c. Do not erect any new structures facing heads of tanks—especially large tanks. It is amazing how far these tanks can travel if they rupture and catch fire.

4. Maintenance and operating procedures should be checked out.

   a. Make certain that the plant has a good maintenance program, including provision for compliance with such code requirements as those for electrical facilities and grounding.

   b. The plant must have satisfactory standard operating procedures for placing a tank on the line.

   c. Purging of tanks should be required before any maintenance or repair work is done.

   d. If a tank is to remain out of service for any appreciable time, it should be filled with water.

5. Whenever a whitish gray fog is seen in the tank area, treat it as LPG until it is definitely known to be something else. More than one person has been killed by driving into one of these fogs or by accidentally striking a spark nearby.

6. Loading and unloading operations should be carefully scrutinized. A number of methods are used for the transfer of LPG to and from tank cars, tanks, and loading or unloading stations. These systems should be designed by professionals, and step-by-

step operating guides must be followed to the letter. Some additional points to watch are listed below.

    a. There must be complete electrical bonding of all components, including the rails supporting tank cars.

    b. Electrical equipment at transfer points must be explosion-proof and properly maintained.

    c. An attendant must be on duty until the connections are broken.

    d. Operating instructions must be posted.

  7. LPG tanks, regardless of size, should never be located near an air conditioning intake (the fan would drag leaking gas into the building) or above operational windows.

B. LPG disasters. LPG fires involving large amounts of gas are fortunately rare. Those that have occurred have been so devastating that every effort should be made to eliminate this type of fire completely. Some examples of LPG fires are provided below.

  1. Decatur, Illinois, July 19, 1974. A 30,000-gallon LPG tank car leaked as a result of a mishap during coupling operations. The leaking gas ignited and, in the ensuing explosion and fires, 7 workmen were killed, 150 people were injured, and approximately 500 freight cars and 2,000 buildings were destroyed or damaged. Dollar losses were in excess of $20 million.

  2. Eagle Pass, Texas, April 30, 1975. An LPG truck ruptured during an accident. In the following explosion and fire, seventeen persons died and others were injured. Buildings were damaged by fire and by flying tank parts that traveled more than 1,000 ft.

IV. Liquefied natural gas (LNG). The use of liquefied natural gas is rapidly increasing because of the ease with which natural gas can be liquefied and stored for vaporization as a supplementary supply during periods of peak demand. Currently, there are a number of large storage locations, and more are on the drawing boards. In addition, there are now tankers of more than 100,000 tons rating plying the oceans with cargoes of pressurized, refrigerated, liquefied natural gas. Still larger tankers are under consideration. The potential hazards should be well thought out before any new large-scale facilities are established. For example:

A. Assume a relatively small tank, 50 ft. in diameter and 30 ft. high. Temperature and pressure are such that the expansion ratio is 600 to 1.

    The tank volume is $\pi r^2 h$ (r = radium and h = height). Therefore, tank volume is 3.14 x 25 x 25 x 30 = 58,865 cu. ft. To find the volume of the potential gas cloud, this figure must be multiplied by the expansion ratio at ambient temperature and pressure. 58,865 x 600 = 35,319,000 cu. ft. This gas cloud will mix with air. The explosive range for LNG is from 5.3 percent to 13.9 percent. If the ratio of gas to air is 7 percent, 504.5 million cu. ft. could reach the explosive range. Allowing for differences of vapor density and for escape of gas on the fringes of the cloud, a more realistic estimate would be 300 million cu. ft. A flammable gas-air mix of this size would cover approximately 230 acres to a height of 30 ft. It should be obvious that large LNG tanks should not be allowed in thickly populated areas.

B. There are two possible methods for unloading tankers: (1) The ships may unload at dockside, preferably in an isolated area; (2) Pipelines may run out to sea and connect with tankers several miles offshore and out of shipping lanes. Of the two methods, only the second appears to the author to be acceptable. Any one of these ships has sufficient gas on board to devastate a large area in the event of accident. (For example, a ship collision could break the thermal insulation. These tankers are floating Thermos jugs.)

C. LNG unloaded from a tanker must be transported to the point of land storage or use. This movement may be by pipeline, railway tank car, or tank truck. Each method must adhere to systems of safeguards prescribed by the U.S. Department of Transportation. Despite these safeguards, there are still some potential hazards that should be checked out.

1. Intrastate shipments. Unless these shipments are made by an interstate carrier, they do not come under DOT supervision. Most states, but not all, have the same requirements as DOT. If there are no state laws governing such shipments, every effort should be made to secure voluntary compliance to DOT standards.

2. Maintenance programs should be in effect and carefully followed through.

3. Routing, particularly of pipelines and tank trucks, should be controlled to minimize exposure to settled areas.

D. Several other potential hazards are the same as those previously noted for LPG and will not be repeated here. A major point of difference in handling the two gases is ventilation of indoor tanks. LNG gases are lighter than air at ordinary ambient temperatures, and leaking LNG tends to rise as it warms up. Therefore, ways must be provided to vent small leaks so that the gas does not rise to higher stories.

E. As with LPG, a moderate volume of LNG mixed with air can cause major explosions and fire. In Cleveland, Ohio, on October 20, 1944, two LNG tanks exploded in sequence and liquefied gas got into the sewer system. As the gas warmed up, it rose, entered buildings and, on finding an ignition source, exploded. Before it was all over, 130 persons died, seventy-nine buildings were destroyed, many additional buildings were damaged, and water mains and electric lines in the area were shattered. Direct property damage was in the millions.

V. Storage facilities. Previous chapters and the earlier part of this chapter have provided an overall look at bulk storage facilities, including common carriers, and their fire problem potential. It is equally important to analyze the smaller types of containers. They are met with far more frequently than bulk containers, and, if wrongly handled, they possess the potential for serious fires on a somewhat smaller scale. Most smaller containers are pressurized cylinders. (Gases may also be stored in cryogenic containers, which are highly specialized, Thermos-like jugs and are not met with too frequently by most fire prevention personnel.) Carelessness in the way cylinders are stored, handled, and used can create serious fire hazard potentials. Some areas of concern are as follows:

A. Cylinder markings.

1. Every cylinder should carry a DOT specification number or be otherwise certified to be in compliance with DOT guidelines for pressure vessels. Some vessels carry numbers from the U.S. Interstate Commerce Commission (ICC), which formerly set specifications.

2. Every cylinder should be clearly marked with the name of the gas it is designed to hold. If the marking is removed, the cylinder must be recertified before reuse.

3. Manufacturers use color coding to identify their cylinders, but different manufacturers use different systems. Do not assume that, because Company A uses green for oxygen cylinders, Company B does also. It is not safe to identify gases by cylinder color only unless the color code actually in use is known.

4. Most cylinders carry red labels, identifying flammable gases, or green labels, identifying nonflammable gases. In certain instances, DOT also requires poison gas or other labeling. Don't assume that the absence of a red label or a poison gas label means the cylinder is safe. Some intrastate operations are not under DOT surveillance and may not require labelling. Know what the gas is.

B. Cylinder changing. Cylinders are designed to contain specific gases. Under no circumstances should a cylinder meant for one gas ever be used to contain another gas.

C. Mechanical injury. Gas cylinders should be protected as necessary to prevent injury, particularly to the valve head.

D. Manifolds. It is quite common for a number of cylinders to be tied into a manifold "pigtail." For example, a large scale welding operation at a fixed location might be supplied by an oxygen manifold and an acetylene manifold. To avoid the possibility of getting the wrong gas in a piped system, each type of gas cylinder has different outlet threads and outlet diameters. Even though a connection to the wrong manifold cannot be made, it is possible to cross thread, and this can be dangerous if a particular mistake is made. The writer remembers a case where an outlet was cross threaded and the manifold valves were open. Gas from cylinders on the line leaked back through the open valves and the cross thread. An explosion occurred that killed the employees in the area, blew out a wall, and shattered windows for blocks.

E. Cylinder holders. Special slings have been developed for moving a number of cylinders at a time. These are the only kind of sling that should be used.

F. Canopies. Highly pressurized cylinders have ruptured as a result of an increase in pressure generated by the heat of the sun. Cylinders stored outdoors should be under canopies.

G. Outdoor pads. Cylinders outdoors should never be stored on the ground. Concrete pads or their equivalent should be used.

H. Cylinder defects. Cylinders should be checked for corrosion and valve defects. Any faulty cylinders found indoors should be removed outdoors. All defective cylinders should be returned to the manufacturer as promptly as possible.

I. Noncompatible gases. Certain gases react with each other. For example, oxygen reacts with flammable gases, and hydrogen reacts with chlorine. Noncompatible gases should preferably not be stored in the same area. If this cannot be done, the following precautions should be observed.

1. There should be gas-tight barriers between the noncompatible gases.

2. Each gas cell should have a separate means of ventilation so that, in the event of leaks, the two gases will not come together.

J. Bulk cylinder storage. Quantities of cylinders should be stored in a detached building or a yard shed. If this cannot be done, cylinders should be in separate storage rooms with access from the outside only.

K. Safety devices. DOT has requirements for various safety devices, depending on the gas and the size of the cylinder. These devices should never be blocked.

L. Storage room heating. Heating should be by steam or hot water.

M. Fire protection. In many instances, automatic sprinklers have, by cooling cylinders not immediately involved in a fire, prevented a serious incident.

N. Ventilation and explosion-proof electrical equipment should be provided as required by the nature of the gas being used.

O. Unless storage sheds have open sides, explosion venting areas are desirable for most locations.

VI. Piping systems. A number of flammable gases are widely used in industry, institutions, commercial establishments, and housing. In most instances, their use involves piping systems, which tend to be overlooked because they are relatively simple to lay out and very common. The flammable gases most commonly conducted through piping systems include natural gas, manufactured gas, and hydrogen. Certain problem areas tend to be overlooked during inspections.

A. Buried piping may not have been installed with a protective coating against corrosion. Such pipe will eventually corrode through and release gas. Pipe may also break because the building settles or because of inadequate shielding from impact forces generated by vehicles overhead. Depending on the porosity of the soil and the presence or absence of water, sanitary, and other connections, leaking gas may travel a long distance to a point of entry and then rise through a building by means of pipe chases or the like. Utility and other lines entering a building should have the spaces between the lines and the building wall properly sealed.

B. Some gases contain appreciable moisture. If the gas piping passes through unheated areas, it must be pitched back to an appropriate condensate drain.

C. Flammable-gas piping should not be buried under buildings.

D. Exposed piping must be protected against mechanical injury.

E. In areas susceptible to earthquakes, Dresser couplings or their equivalent should be provided near the point of entry to the building and inside the building as necessary.

F. No gas piping should ever be installed under a foundation wall or footing.

Flammable Liquids and Gases

G. All gas piping supplied from outside storage or distribution systems should have an accessible outside control valve.

H. When a service is eliminated permanently or temporarily, the lines should be capped—outside the building if possible. Do not depend entirely on the gas valve. (In Cleveland some years ago, a gas line was shut off in the street while an oven was being modified. The valve leaked, and gas came through the open-ended pipe and was ignited by a welding torch. The building was heavily damaged. In Massachusetts, an old textile mill was being renovated. Workmen broke through a false wall inside the original wall. A pickax striking an open-end, abandoned gas line generated sparks that ignited a gas-air mix in the concealed space. Three workers were killed or seriously injured by the explosion.)

I. In systems that use gas at lower pressure than is supplied, make certain regulator valves are carefully checked.

J. All exposed piping should be regularly inspected and maintained.

K. Hydrogen, acetylene, and some other flammable-gas piping systems are normally provided with hydraulic arrestors at key points to prevent flashbacks in the lines. These arrestors must be maintained in service.

VII. Acetylene generation. One of the most commonly used industrial gases is acetylene. The gas has an unusually wide explosive range, tends to be unstable, and, in contact with copper or silver, forms compounds that can easily be detonated. Therefore, acetylene generation plants must be very carefully designed by professionals in this field. It is equally important that no alterations be made in plant or equipment without approval of qualified personnel.

There are several acceptable ways of manufacturing acetylene. Each has its own set of prescribed operating procedures. Operating personnel should be trained in the correct procedures for their particular operation. In addition to operating procedures, certain other safety considerations are of concern to fire prevention personnel. Some of the most important are listed below.

A. Calcium carbide should always be stored in unopened containers in a noncombustible or fire resistive room out of reach of water.

B. It is essential that carbide storage areas have ceiling ventilation and floor level fresh air intakes and that these facilities be maintained at all times.

C. Cylinders to be filled with acetylene must be stored upright and handled gently. Such cylinders contain a porous mineral filler and acetone. Mishandling may damage the filler and convert a stable gas setup to an unstable situation.

D. Room heating should be by steam or hot water.

E. Generator rooms should be of damage-limiting construction.

F. Generators should be completely overhauled several times a year.

G. Because of the potential hazards, there must be exceptionally good inspection and maintenance programs.

VIII. Cutting and welding.

A. Hazards. Improper cutting and welding procedures have been responsible for many large fires. This situation is probably due in part to a lack of understanding of the ways in which these operations can start fires. Principal origins of fires associated with cutting and welding are listed below.

1. The torch itself.
2. Heat conduction through the metal being worked on.
3. Molten slag and/or metal from the cut or weld.
4. Sparks.
5. Leaking acetylene.
6. Regulator failure.

B. Safety procedures. By and large, these fires can be eliminated if the following procedures become standard practice.

1. Maintain regulators and allied cylinder equipment in operating order. This will prevent backflow of oxygen into the fuel gas supply, flashbacks into the fuel supply, and excessive oxygen backpressure.

2. Whenever a welding outfit is shut off and left unattended, close the valves at the cylinders. This will prevent gas leakage in the event the hose is damaged. (Although a hose should not be left on the floor, it is inevitable that at times it will be. More than one hose line with pressure on has been cut by a forklift.)

3. Use only standard welding hoses—green for oxygen, red for fuel. Connections must be tight and made without the use of pipe compound, white lead, or grease.

4. No cutting or welding should be permitted where an open flame would not be allowed.

5. All cutting and welding should be done on a permit basis. Permits should be issued only by personnel with fire safety responsibility.

6. Areas where cutting or welding is to be done should be

checked in advance for possible holes or cracks through which molten metal or sparks could travel to another area. Cover such openings with sheet metal, asbestos board, or their equal before operations are started.

7. Necessary cutting or welding on equipment in a hazardous location should, if possible, be moved to a safe area. Otherwise, shut down the hazardous process and remove or protect the hazardous materials. If cutting or welding must be done where combustible construction and/or combustible materials are present, there should always be a fire watch with appropriate extinguishers. The fire watch should remain in the area for thirty to forty-five minutes after the operation has been completed to make sure there are no smoldering fires.

8. Containers that have contained flammables should not be cut or welded until the containers have been emptied and purged as required by the nature of the former contents.

9. Bulk storage of acetylene cylinders should be kept away from oxygen storage.

10. No cutting or welding should be permitted in a building requiring sprinkler protection when the sprinklers are out of service.

Summation

Like flammable liquids, flammable gases are widely used, and there is no way to detail all the areas of potential fire hazard. However, an inspector need not know such detail provided he has a good understanding of the basic gas laws and the basic properties of flammable gases. With this foundation, and applying such examples as those previously cited, he or she can generally achieve fire safety for flammable gases by applying common sense. Where unusual conditions requiring special treatment do exist, an inspector should not hesitate to use reference materials and request advice from qualified experts in the appropriate fields.

As previously noted, the Department of Transportation has published warning labels for flammable liquids and gases and for other hazardous materials that are shipped interstate or to foreign nations. Charts of currently used labels and placards are shown in figures 16, 17, and 18.

Fig. 16 Newly authorized hazardous materials warning labels
(see front end sheets for illustration)

GENERAL GUIDELINES ON USE OF LABELS

1. Shipper must furnish and attach appropriate label(s) to each package of hazardous material offered for shipment unless exempted from labeling requirements. (Ref. Title 49, CFR, Sec. 173.404(a)).

2. If the material in a package has more than one hazard classification, one of which is Class A explosives, Class A poison, or Radioactive Materials, the package must be labeled for <u>each</u> hazard. (Ref. Title 49, CFR, Sec. 173.402(b)).

3. When two or more hazardous materials of different classes are packed within the same packaging or outer enclosure, the outside of the package must be labeled for <u>each</u> material involved. (Ref. Title 49, CFR, Sec. 173.403(a)).

4. Radioactive materials requiring labeling, must be labeled on two opposite sides of the package. (Ref. Title 49, CFR, Sec. 173.402(a) (10)).

5. Labels must not be applied to a package containing only material which is not subject to Parts 170-189 of this subchapter or which is exempted therefrom. However, this paragraph does not prohibit the use of labels required for purposes of import or export shipments or required by 14 CFR 103.13 of the Federal Aviation Regulations on packages destined for transportation by air. (Ref. Title 49, CFR, Sec. 173.404(b)).

EXPORT AND IMPORT SHIPMENTS

EXPORT SHIPMENTS – Exporters are advised that shipments by water or air to foreign destinations may be rejected for transportation if they bear warning labels other than those illustrated in this chart, since many countries are implementing labeling requirements conforming to the United Nations Recommendations.

IMPORT SHIPMENTS – Labels having the same size, color and symbols as prescribed by Title 49 affixed to packages in another country are authorized for shipments in the USA. They may contain inscriptions required by the country of origin. (Ref. Title 49, CFR, Sec. 173.404(f)).

NOTE: The section numbers and references shown above are found in the Code of Federal Regulations, CFR Title 49-Transportation, Parts 100-199 and CFR Title 14-Aeronautics and Space, Parts 60-199. The same references may be found in the following publications which also contain a reprint of Title 49 CFR Parts 100-199:

R.M. Graziano's Tariff (Bureau of Explosives)
ATA (American Trucking Association) Dangerous Articles Tariff

Fig. 17. Truck placarding chart
(see back end sheet)

Fig. 18. Hazardous materials placards for rail cars

KEY TO USE OF RAIL PLACARDS
Ref: Title 49, CFR, PART 174

RAIL CARS ARE REQUIRED TO BE PLACARDED AS DESCRIBED BELOW. PLACARDS AS REQUIRED, MUST BE APPLIED TO BOTH ENDS AND BOTH SIDES OF THE CAR. FOR ADDITIONAL INFORMATION, SEE SECTIONS 174.540 THRU 174.557.

1 DANGEROUS PLACARD—Used on:

    a. Cars loaded with bulk shipments and tank cars containing flammable liquids or solids, oxidizing materials, acids and corrosive liquids, Class B poisons, compressed flammable gases and compressed nonflammable gases. (Ref. Sec. 174.541(a)(2), (3) and 174.552)

    b. Other than tank cars containing one or more labeled packages of flammable liquids or solids, oxidizing materials, corrosive liquids, Class B poisons and Class B explosives. (Ref. Sec. 174.541(a)(1) and (4))

2 DANGEROUS—EMPTY PLACARD—Used on tank cars after material requiring dangerous placard has been unloaded to warn personnel of the potential hazards. (Ref. Sec. 174.562(b) and 174.563)

3 CAUTION—RESIDUAL PHOSPHORUS—EMPTY PLACARD—Used to warn personnel that tank cars last contained white or yellow phosphorus. (Ref. Sec. 174.541(c) and 174.555)

4 EXPLOSIVE PLACARD—Used on cars containing any quantity of Class A explosives. (Ref. Sec. 174.540, 174.541(a)(5) and 174.550)

5 DANGEROUS—RADIOACTIVE MATERIAL PLACARD—Used on cars containing any quantity of "radioactive yellow III" and to carload lots of certain other radioactive materials. (Ref. Sec. 174.541(b), 174.553, 173.392 and 173.393(j))

6 FLAMMABLE POISON GAS PLACARD—Used on Class 105A-W tank cars containing hydrocyanic acid. (Ref. Sec. 174-542(b) and 174.556)

7 FLAMMABLE POISON GAS—EMPTY PLACARD—Used on tank cars after a materials requiring a flammable poison gas placard has been unloaded. (Ref. Sec. 174.562(d) and 174.563(d))

8 POISON GAS PLACARD—Used on tank cars containing Class A poisons except hydrocyanic acid. (Ref. Sec. 174.542(c) and 174.557)

9 POISON GAS—EMPTY PLACARD—Used on empty tank cars that last contained a Class A poison other than hydrocyanic acid. (Ref. Sec. 174.562(e) and 174.563(f))

10 POISON GAS PLACARD—Used on cars other than tank cars that contain packages or containers bearing a "Poison Gas" label. (Ref. Sec. 174.542(a)(1)(2) and 174.551)

11 DANGER—FUMIGATED OR TREATED PLACARD—Used on cars containing lading which has been fumigated or treated with poisonous liquids, solids or gases. (Ref. Sec. 174.579)

☆ U.S. GOVERNMENT PRINTING OFFICE : 1973—726-831/525 3-1

Fig. 18. (continued)

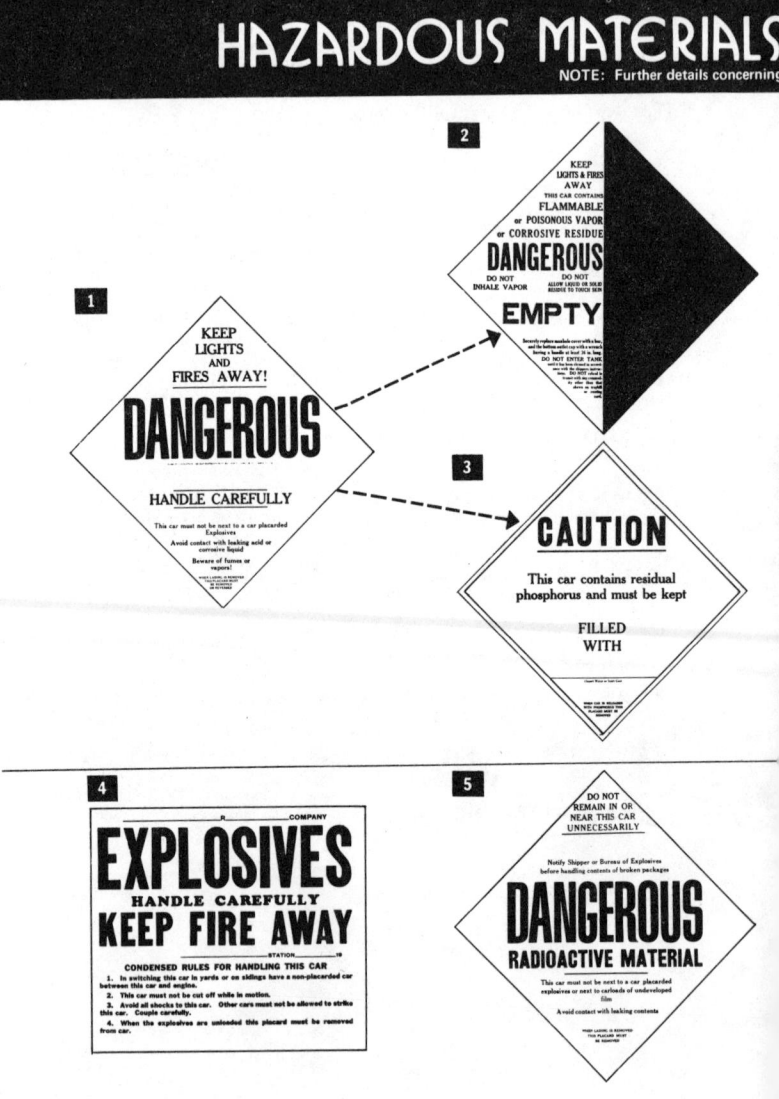

CARDS FOR RAIL CARS

cards are included on the back cover.

6
DO NOT REMAIN ON OR NEAR THIS CAR UNNECESSARILY

Lading must not be transferred en route under any conditions before shipper and Bureau of Explosives are notified.

FLAMMABLE POISON GAS

NAME OF CONTENTS

This car must not be next to a car placarded "Explosives". Beware of liquid and of gas leaking from tank or fittings.

WHEN LADING IS REMOVED THIS PLACARD MUST BE REVERSED

(FOR TANK CARS)

7
KEEP LIGHTS AND FIRES AWAY

THIS CAR CONTAINS FLAMMABLE POISON GAS OR RESIDUE

DANGEROUS

DO NOT INHALE GAS — DO NOT ALLOW LIQUID OR SOLID RESIDUE TO TOUCH SKIN

EMPTY

KEEP MANWAY BONNET COVER SECURELY CLOSED. DO NOT ENTER TANK UNTIL IT HAS BEEN CLEANED IN ACCORDANCE WITH THE SHIPPERS INSTRUCTIONS

DO NOT RELOAD IN TRANSIT

8
DO NOT REMAIN ON OR NEAR THIS CAR UNNECESSARILY

Lading must not be transferred en route under any conditions before shipper Bureau of Explosives are notified.

POISON GAS

NAME OF CONTENTS

This car must not be next to a car placarded "Explosives". Beware of liquid and of gas leaking from tank or fittings.

WHEN LADING IS REMOVED THIS PLACARD MUST BE REVERSED

(FOR TANK CARS)

9
KEEP LIGHTS AND FIRES AWAY

THIS CAR CONTAINS POISON GAS OR RESIDUE

DANGEROUS

DO NOT INHALE GAS — DO NOT ALLOW LIQUID OR SOLID RESIDUE TO TOUCH SKIN

EMPTY

KEEP MANWAY BONNET COVER SECURELY CLOSED. DO NOT ENTER TANK UNTIL IT HAS BEEN CLEANED IN ACCORDANCE WITH THE SHIPPERS INSTRUCTIONS

DO NOT RELOAD IN TRANSIT

10
CAUTION
This Car Contains

POISON GAS

Beware of Fumes from Leaking Packages.

FOR OTHER THAN TANK CARS)

11
DANGER
The lading of this car has been

FUMIGATED or TREATED

with

(Name of poisonous liquid, solid, or gas)

BEFORE UNLOADING, open both doors and DO NOT ENTER until car is free of gas. REMOVE ALL POISONOUS MATERIAL before release of empty car.

chapter 10
Special Chemical Hazards

It would be highly desirable to eliminate the use of all hazardous materials—if it were possible to do so and still maintain the present standard of living. However, it can be positively shown that eliminating hazardous materials to any great extent would set back the clock 100 years or more. The result would be shorter life spans, much longer working hours, and the elimination of many things that make life easier and more comfortable.

Within the past seventy-five years, we have learned to live with such hazards as oxygen—for example, in medicine and space flight; gasoline in transportation; magnesium and aluminum alloys in vehicle bodies and wheels, kitchen utensils, and the like; natural gas for heat and power; radioactive materials for nuclear power, medicine, and many industrial uses; and synthetic fibers in clothing. These are but a few of the ways in which hazardous materials, used under controlled conditions, benefit mankind. True, failure of controls can result in a disaster, but the overall benefits have far outweighed the failures.

The hazards of chemicals such as those named above are well known, as are the appropriate means of hazard control. Not so well known are the hazards presented by certain special chemical groups. These chemicals are very necessary to the efficient operation of a number of modern industrial processes, and their lack in some cases would mean the elimination of very desirable products. It is therefore important that fire control personnel learn some basic facts both about the hazards presented by these chemicals and about how to control the hazards.

Various methods are used to group special hazard chemicals. From a practical standpoint, it would appear desirable to use the same arrangement as that used by many fire departments, and that is the method followed here. Main group headings are as follows:

1. Air- and water-reactive chemicals
2. Oxidizing agents[1]
3. Corrosive chemicals
4. Toxic chemicals
5. Explosives
6. Cryogenics
7. Unstable chemicals
8. Radioactive materials[2]

There is no single publication, nor is there any single individual, with all the answers to special chemical fire problems. No one expects a fire prevention official—whether inspector, engineer, chemist, or fire marshal—to know all the solutions or, if there is more than one solution, to always know which is economically the most practical. However, concerned individuals should know where to look for help in finding answers to problems beyond their areas of knowledge. It is desirable that such individuals own or have access to a basic library that can provide initial answers pending further information from a qualified expert.

Air- and Water-Reactive Chemicals

A number of groups of chemicals undergo reactions of varying intensities when in contact with air, moisture, or water. Such reactions may cause spontaneous ignition, generate sufficient heat to ignite combustibles, or be very violent. With some chemicals, primary hazard of these materials is corrosiveness or toxicity, and information on these substances has been omitted from this section.

I. Major groupings and typical locations. Cited below are only some

---

[1] Included in this grouping are those halogens that are, in effect, vigorous oxidizers.

[2] To describe the basic fire problems associated with radioactive materials would require more space than can be provided in a general purpose text. Information on this group has therefore been omitted.

Special Chemical Hazards 187

of the groups of hazardous chemicals that react strongly with air and/or water. However, the list does indicate that the usage of such materials is widespread and warrants attention from the fire prevention standpoint.

A. Alkyl boranes. These substances may be solids, liquids, or gases, and they range from ordinary hazard to extremely hazardous. An example of the latter is diborane, a highly flammable gas with a wide explosive range. Diborane also is toxic and very reactive. When the name borane appears on a container, don't guess. Look it up. A major area of use for these substances is as rocket propellants.

B. Aluminum alkyls. These alkyls are liquids that are often shipped dissolved in a hydrocarbon such as benzene, toluene, or hexane. They react vigorously with water, oxidizers, halogens, alcohols, and other chemicals. Aluminum alkyls are principally used as rocket propellants.

C. Acid anhydrides. Acid anhydrides react vigorously with water to regenerate acids. Organic anhydrides are combustible and generally have relatively low flash points. Inorganic anhydrides are not combustible but may, in contact with combustibles, generate enough heat to ignite them. Anhydrides are used in dehydration processes.

D. Carbides. Virtually all carbides react with water, some explosively. Some, such as calcium carbide, react with water to form acetylene. Others, such as aluminum carbide, react to form methane. Sodium and potassium carbides react explosively with water. On the other hand, silicon carbide is carborundum, which of course is nonhazardous.

E. Chlorides. These chemicals form a heterogeneous group. They include table salt; the antifreeze material calcium chloride; aluminum chloride, which reacts with water to form hydrogen chloride; and acetyl chloride, which, when heated, liberates the poison gas phosgene plus hydrogen chloride.

F. Hydrides. In contact with water, metallic hydrides react vigorously or even violently to form hydrogen—which, in the presence of an ignition source, can explode with devastating force. Sodium and potassium hydrides react vigorously with moist air. The writer remembers an explosion some twenty-five years ago caused by sodium hydride in contact with moist air. Three men were literally blown apart, and a section of a building was demolished. Hydrides are used for rust desealing and in the pharmaceutical field.

G. Alkali metals. All the most commonly used alkali metals—sodium, potassium, lithium, and calcium—react with water (calcium less vigorously than the others). Certain rare metals also react vigorously with water. The reactions, in general, produce hydrogen and caustics. It is important to remember that the more reactive alkali metals—and, to a lesser extent, other water-reactive materials—should never come in contact with a concrete floor that may contain water. Alkali metals are used in heat exchangers, at nuclear reactors, in the chemical industry, and in certain other operations.

H. Phosphorus and its compounds. These chemicals are normally found in plants manufacturing soaps, detergents, or fertilizers. White and yellow phosphorus react with air to form oxides. Red phosphorus does not oxidize directly but, when heated, reverts to white phosphorus. Pure phosphorus is normally stored under water. In contrast, many phosphorus compounds react with water (some burn) and give off such toxic and flammable gases as hydrogen sulfide.

I. Silanes. These materials consist of combinations of hydrogen, silicon, and other elements. To various degrees, silanes are susceptible to spontaneous ignition in air; some are highly flammable, and many are seriously corrosive. These materials are used in rubber- and resin-manufacturing operations and in doping agents.

II. Fire protection. Special measures are needed to provide proper fire protection for air- and water-reactive chemicals. Many of these substances react with water to generate fire, release dangerously toxic and/or flammable gases, or form strong acids or caustics. In air, reactive chemicals may burn spontaneously, and some of the resulting oxides are capable of generating enough heat in contact with water to ignite other combustibles. Some of the precautionary fire prevention and protection measures that may be necessary are listed below.

A. Water-reactive chemicals.

1. Store in watertight and airtight containers that are kept off the floor by skids.
2. Preferably, store in fire resistive or nonflammable buildings.
3. Limit the quantity of hazardous chemicals stored.
4. If large amounts of water-reactive materials are needed, split up the storage into small lots in buildings that will not expose other important property.
5. Appropriate inert materials should be provided for fire extinguishment.
6. Do not store materials with which the hazardous materials may react in the same area with the hazardous materials. In general, do not store combustibles in the area either.

7. If reactions with water are not likely to be violent, limited combustibles in the same area and a combustible building with sprinkler protection are acceptable. If reactions are likely to be violent, combustible storage or construction is not acceptable, and sprinkler protection should not be provided. In addition, buildings should be posted against the use of water.

8. Building heating systems should be of the indirect type—for example, dry hot air.

B. Air-reactive chemicals. These materials should be stored under water or inert gas. Large volumes of water (sprinkler and/or hose streams) should be available for fire extinguishment. Wet down hazardous chemicals should be covered by inert material and promptly removed from the premises for disposal. Failure to do this will result in reignition when the material dries out.

C. General.

1. In the event of fire, many air- and water-reactive chemicals pose serious problems for responding fire departments. There should therefore be close coordination between plants using these materials and fire departments.

2. Plants having appreciable quantities of air- and/or water-reactive chemicals on the premises should have trained, qualified personnel to monitor the storage and use of these chemicals and to take the necessary actions in the event of an emergency.

III. Manufacturing. There are too many potential ways to use hazardous chemicals in manufacturing operations to list them all here. For all such activities, however, it should be a standing rule to keep quantities to an absolute minimum and to post a list of essential safety precautions and enforce them.

IV. Summation. It is important to note that many of the materials in this grouping not only react vigorously or violently with air and/or water, but also are highly combustible. Others produce dangerously toxic gases during reactions or fires or produce acids or caustics as a result of their reactions. Quite a few are hazardous in more than one way.

It is not possible for anyone to remember all the chemical details of the thousands of materials in the various hazardous categories. However it is possible for a person to learn the basic groupings with which he or she is likely to come in contact and where these materials are likely to be found. With this background, the availability of good reference sources, and experience, it is possible to develop sound judgment that will result in a high degree of safety.

Oxidizing Agents

The process of oxidation ranges from a very slow process such as rusting to extremely rapid reactions that can result in a fast-moving fire or a major explosion. It is well to be aware that the same materials under different conditions may produce widely differing types of reactions. Illustrations of such variations are provided below.

1. When the leaves of hardwood trees stop producing chlorophyll, one of the results is that the leaves start to oxidize. Under normal conditions, this process plays a major part in the fall spectacles of color. However, when the right conditions of temperature and humidity occur at the right time, an entire valley or hillside can in the presence of an ignition source erupt into a sheet of flame in minutes or even seconds. Both situations of course involve oxidation.

2. Sheet aluminum is used for house siding and cooking utensils. In both instances, exposed surfaces oxidize in a harmless manner. In contrast, aluminum powder in air in the presence of an ignition source is quite capable of generating pressures of 60 psi or more. Such pressures can and have demolished or wrecked buildings that were not adequately vented.

The above examples should make it obvious that oxidizing chemicals need to be considered from the standpoint, not only of their oxidizing capabilities, but also of how and where they are used.

Many oxidizing chemicals also have other important hazard properties, and these materials will be cited in other sections of this chapter. This section will deal chiefly with solid oxidizers, oxygen, and the halogens.

I. Areas of use. When the subject of industrial and institutional use of oxidizers is brought up, most people tend to think of liquid or gaseous oxygen. Direct oxygen use is highly important, but it is far from being the only important oxidation process. Listed below are some operations where oxidizers are present and the potential for damage is frequently overlooked. It should be noted that many of the operations listed are not ordinarily thought of as involving oxidizers, at least in a form that creates a potential fire problem.

    A. Fertilizer manufacture and storage.
    B. Manufacture and storage of drugs.
    C. Silvering mirrors.
    D. Plastics fabrication.
    E. Certain forms of electroplating.
    F. Manufacture and storage of bleaching agents.

Special Chemical Hazards

    G. Water purification.
    H. Chemical warehousing.
II. Some common solid oxidizers.
    A. Bromates. All are powerful oxidizers, and some emit toxic fumes while decomposing.
    B. Chlorates. All are powerful oxidizers.
    C. Hypochlorites. In addition to their oxidizing capability, they release chlorine when exposed to moisture.
    D. Nitrates. These strong oxidizers can cause violent explosions. Nitrates also release toxic oxides of nitrogen when heated.
    E. Permanganates and persulfates are among the other oxidizers used.

There are so many oxidizing agents in use that they cannot all be described in detail except in a chemical handbook. All of them, however, form part of a grouping such as those listed above, and the basic characteristics of each group can be studied. Fire prevention personnel need study an individual chemical within a group only if the substance has one or more characteristics that deviate sharply from those of the group as a whole. For example, most commonly used nitrates have a reactivity rating of 1, and precautionary measures should be set up accordingly. However, ammonium nitrate has a reactivity rating of 3, which means an explosion is possible. When dealing with nitrates, therefore, a fire prevention officer should take precautions against the possiblity of fast-spreading fires resulting from the release of oxygen. In dealing with ammonium nitrate, he or she must also pay special attention to minimizing the possibility of explosions.

III. Reactions of oxidizers to fire or heat. Fire, of course, makes for fast release of oxygen, but it also is entirely possible for severe reactions to result from a relatively low order of heat. For example, sulfur melts at 234° F. and ammonium nitrate at 337° F. Wrongly located, these two materials could melt and flow together as the result of heat from a nearby fire or some other source. At that point, with an ignition source or possibly with no ignition source, there would be an explosion, and the warehouse would disappear. This possibility and others like it show why combustible materials should not be stored in the same area with oxidizers. (Many oxidizers have a low melting point and so do a number of combustibles.) Similarly, railroad freight cars containing oxidizers are normally hauled in a different part of the train from those carrying combustibles so as to avoid a disastrous chemical reaction in the event of a derailment.

IV. Elementary safety precautions.

A. Oxidizing agents should be stored in fire resistive noncombustible buildings which, if an explosion hazard is present, should be suitably vented. If such substances are stored outside, as is sometimes done, they should preferably be on a concrete pad; vegetation should be controlled, and, if an explosion hazard may be present, the oxidizers should be located behind berms or other barriers.

B. Protection systems should be provided as required for the storage location. In general, these should be sprinkler systems.

C. Oxidizers should always be separated by distance or barriers from combustibles and reactive materials.

D. In some instances, special electrical equipment may be needed.

E. Containers must be in a cool, dry location and protected against mechanical injury.

F. Means must be available for the prompt removal of spills or contaminated material.

G. Plants or institutions handling oxidizers in hazardous quantities should have qualified personnel to "ride herd" on the storage and use of oxidizers.

H. In areas where oxidizers are being used, inspectors should be alert to any special insurance requirements and how well these are being carried out.

V. Ammonium nitrate. This substance is probably the most common oxidizer in use. In fertilizer operations alone, it is found in large quantities in manufacturing plants, in warehouses, and in transit. In more moderate quantities, it is found on farms, in plant nurseries, in stores, and in many residences. Wherever it is, ammonium nitrate is an oxidizer. It is important to recognize this, because many people believe that fertilizer represents no fire problem regardless of its amount or location. This is simply not so.

It is true that ammonium nitrate does not burn. It is also true that, exposed to heat from a fire, ammonium nitrate liberates oxygen and speeds up the fire. Furthermore, devastating explosions can take place if a considerable amount is present and it becomes contaminated with organic materials. For example:

A. Texas City, Texas, 1947. A shipboard explosion killed more than 500 people, virtually destroyed the waterfront, and shattered the Texas City Fire Department.

B. Germany, about 1923. A plant explosion killed about 800 workmen.

C. Arkansas, 1960. A train wreck killed five and destroyed or badly damaged the train and nearby facilities.

Special Chemical Hazards 193

VI. Oxygen. Oxygen is widely used in both gaseous and liquid forms. Liquid oxygen is a cryogenic liquid and is detailed under that heading later in this chapter. Gaseous oxygen has been used for a long time, and the appropriate general safeguards are well known. Unfortunately, there is a tendency to take familiar objects for granted, and as a result some elementary safeguards may be overlooked. These safeguards include the following:

A. Store cylinders upright. If a cylinder in this position failed at the valve head during a fire, the propulsion force would be downwards. A valve head failure could cause a horizontal cylinder to become a projectile that could travel from $\frac{1}{4}$ to $\frac{1}{2}$ mile.

B. Chain cylinders in place. This prevents them from being knocked down and possibly snapping off a valve head.

C. When cylinders are not in use, screw on valve caps tightly to prevent mechanical injury.

D. Take care that certain valves and regulators do not become contaminated with oil or grease.

E. If cylinders are stored against a wall or in a mass group, be sure that safety pressure release devices are not blocked. Otherwise, in event of fire an explosive cylinder rupture might occur.

F. Never store oxygen cylinders in the same area as combustible gases, such as acetylene, or other highly combustible materials. If contiguous storage cannot be avoided and quantities are more than nominal, consideration should be given to gas-tight barriers.

G. Oxygen piping systems require special design and the use of approved fittings. They also require exceptional maintenance. Any extensions or replacements must be of the same quality as the original system.

VII. Manufacturing operations. Standards have been set for the manufacture of oxidizers, but the standards for the different kinds of products differ considerably. Plants should have available both the appropriate standards and the recommendations of their insurance carriers and have the organization necessary to make certain these safeguards are complied with.

VIII. Overhaul. If a fire has occurred and oxidizers have spread near combustible construction or materials, the building should not be reoccupied until questionable areas have been decontaminated by experts. Oxidizers dissolved in water may impregnate wood and other combustible materials. The hazard occurs when the impregnated material dries out.

IX. Heat treating and salt baths. Nitrates are the chemicals most commonly used for molten salt baths. They are often used in combination with other oxidizers or with chemicals for descaling, coloring, cleaning, or other purposes. In addition, nitrate baths may stand near other molten baths employing such materials as cyanides, caustics, or chlorides. All of these substances must be kept from mixing with each other. (Some may be stored together, but doing so requires accurate knowledge of the characteristics of each material. This is no place for guesswork.)

In checking out molten baths in heat treating plants, it is vital to ascertain what materials are being used and that they are being used properly. All such baths should have well-maintained temperature controls.

Preferably, the operations should be carried out in a fire resistive or noncombustible building, but sometimes this is not possible. Where construction is in part combustible and/or there are combustible contents, molten salt baths should be shielded from water discharged from sprinklers or from hose streams. It is better to leave by the stairs than through the wall.

If a molten salt bath is malfunctioning, the following actions should be carried out to prevent a fire or explosion.

A. Shut off the heat and remove the work.

B. Evacuate personnel from the area.

C. Make sure medical attention is available: there will almost certainly be toxic fumes.

D. If the bath temperature continues to rise, alert the proper authorities and withdraw. On the way out, ventilate the building, and shut off gas, oil, and electric lines.

X. Halogens. The common halogens are fluorine, chlorine, bromine, and iodine. The first two are gases, bromine is a liquid, and iodine is a solid at ordinary ambient temperatures. All are toxic, and all are oxidizers.

A. Uses.

1. Fluorine is used in masonry-cutting torches, in the manufacturing of fluorocarbons, at missile sites, and elsewhere.

2. Chlorine is used at water treatment plants, in bleacheries, in the manufacture of chlorinated hydrocarbons, and in other operations.

3. Bromine is used in the manufacture of dyes, inks, antiknock fuels, in metal refining, and in other processes.

4. Iodine is used in dyes, pharmaceuticals, and paint pigments.

Special Chemical Hazards

5. Halogens are oxidizers in the sense that they react with many materials in the same sense that oxygen does. That some of them can be combined with other elements to form extinguishing agents does not change this fact.

B. Hazards. Spilled, leaking, or burning halogens present a number of hazards, among which are the following:

1. Fluorine reacts vigorously with most materials, in many instances with sufficient heat to cause fire. It may be said that, in general, a fluorine leak will become a fluorine fire.

2. Chlorine leaks in damp areas eat through metal, possibly releasing highly flammable material.

3. Chlorine may react explosively with hydrogen, methane, ether, and certain other gases.

4. Bromine in contact with certain hydrocarbons generates enough heat to start a fire.

5. A mixture of iodine crystals and certain finely ground combustibles can be detonated.

C. Special precautions. Halogens and some of their compounds present many unusual problems to the fire fighting service. Because of this, and because all four are toxic besides being oxidizers, plants using these materials in any quantity should closely coordinate their safety programs with the responding fire department and with the transportation agency making deliveries.

XI. Summation. Oxidizing chemicals or agents present a number of hazards. However, these hazards can safely be lived with provided there is—

A. basic knowledge of the potential hazards of those oxidizers actually used within the area of responsibility;

B. an awareness of where more detailed information can be obtained as needed;

C. coordination with other agencies or groups having an interest.

Corrosive Chemicals

For the purposes of this chapter, corrosive chemicals are defined as those chemicals whose primary hazard under fire conditions lies in their potential destructive effect on living tissue and on certain inanimate materials. Chemicals that are corrosive but whose primary hazard lies elsewhere are excluded.

As defined above, corrosive chemicals for all practical purposes are liquids or solids and include only acids and alkalis. Of these, some have played major roles in loss of life and property and serious injuries during fires. Others have had little or no effect on the overall picture. It pays to know the difference.

I. Acids. Many types of acids are used throughout industry and elsewhere, but only a few that are commercially important pose any fire problem. They are hydrochloric acid (HCl), nitric acid HNO_3), and sulfuric acid (H_2SO_4). To this list should perhaps be added hydrofluoric acid (HF) which is not very widely used, but which can present a major health hazard.

In any discussion of acids, it is important that the difference between the terms *concentration* and *strength* be clearly understood. Concentration rates the amount of water in which the acid is dissolved. Strength measures the amount of ionization in any given solution. For example, boric acid ionizes very little and sulfuric acid a great deal. Accordingly, 5 percent boric acid is little more than an eye wash, while 5 percent sulfuric acid will burn through human tissue.

A. Hazards. Small quantities of these acids, such as are found in well-run laboratories, should not be a fire problem. But moderate to large quantities of strong acids are found in many locations. If acid liberation should occur in these areas, the most important hazards would be as follows:

1. Appreciable contact of living tissue with even moderate concentrations of strong acids can cause permanent, painful disfigurement or even death.

2. During a fire, liberated nitric acid gives off nitrogen dioxide, which is highly toxic. Similarly, hydrochloric acid gives off hydrogen chloride fumes.

3. Acids are frequently stored in glass carboys or in glass-lined metal drums. Rough handling can break the carboys or the glass lining. Any strong acid, when exposed to sufficient heat, generates enough pressure to break a container.

4. If a concentrated acid is hit by a hose stream, a violent eruption will occur.

5. One hundred percent sulfuric acid is sometimes stored in special steel drums. If the acid absorbs any moisture, hydrogen forms.

6. Liberated nitric acid can form explosives by nitrating finely divided organic material.

7. Either nitric or sulfuric acid can ignite certain finely divided materials. Both can eat through wood floors.

8. All inorganic acids attack metals. It is therefore vitally important to check on what is stored in the same area with them. The

Special Chemical Hazards 197

author once saw a warehouse (concrete and sprinklered) that contained glass carboys of hydrochloric, nitric, and sulfuric acids plus thousands of gallons of flammable liquids in metal containers, oxygen and acetylene cylinders, copper cyanide in fiberboard drums, and other items. If a minor fire had broken one or more acid carboys, an uncontrollable inferno would have developed in short order because of the acid's ability to eat through wood and metal. The acids were moved out the same day, as were certain other items.

9. Acid piping should never run above a water line or a sanitary sewer. Leaking acid can destroy piping, including that for sprinklers.

10. Never install acid piping over human occupancy.

B. Uses. Areas where strong acids are found in moderate to large quantities include, but are not limited to, the following:

 1. Metal cleaning and plating.
 2. Steel and wire mills.
 3. Transportation vehicles.
 4. Chemical warehouses.

II. Alkalis. Like acids, alkalis are used in many areas and differ in their effects on living tissue and certain other materials. Of the alkalis commonly used, the two of most concern from a fire safety standpoint are sodium hydroxide (caustic soda) and potassium hydroxide (caustic potash). Although not a true alkali, calcium oxide (quicklime) reacts in the same manner as true caustics.

A. Hazards. Caustics seldom are directly involved in starting fires or in greatly extending fire spread, but they are or should be of concern to fire prevention personnel because of their potential life hazard in fire situations. Industrial caustics are often found in open pits or in various types of storage containers. In areas where moderate to large quantities of caustics are used in the open or may be liberated, the most important hazards are as follows:

1. Caustics destroy tissue, and healing of serious alkali burns is extremely difficult. A severe degree of contact often means death.

2. Dry caustics react violently with water and generate heat that may be sufficient to ignite nearby combustibles. There is also, of course, a danger from spattering.

3. Strong caustics in contact with aluminum or zinc generate hydrogen.

B. Uses. Operations where strong caustics are found in moderate to large quantities include, but are not limited to, the following:

1. Metal cleaning and plating.
2. Oil refining.
3. Pulp and paper mills.
4. Textile manufacturing.
5. Soap manufacturing.
6. Rubber reclaiming.
7. Chemical warehousing.
8. Transportation.

III. Safety measures. Plants where large amounts of acids or alkalis are present should be prepared to take certain emergency measures in the event of a spill or fire. Of the actions listed below, some would normally be performed by the responsible fire department.

A. Accidental Spills. Emergency actions should include the following:
1. Acids.
    a. Ventilation.
    b. Provision of breathing apparatus.
    c. Provision of suitable materials for neutralizing or absorbing the spilled acid.
    d. Provision of large volumes of water to dilute and disperse the acid where this can be done with no additional damage.
    e. Provision of suitable protective clothing.
    f. Provision of safety showers and medical aid for personnel.
2. Alkalis.
    a. Provision of suitable neutralizing material.
    b. Provision of large volumes of water to dissipate heat and dilute alkali in major spills where corrosion is not a problem.
    c. Provision of protective clothing and essential medical aid.

B. Fires. In addition to direct firefighting capability, there is a need for advance planning to assure safety to life and property in the event of fire. Some of the advance actions that can be taken are as follows:
1. Open vats should have covers that are arranged to close in event of fire. Guard rails are no protection for personnel in smoke-obscured areas.
2. If chemical tanks are elevated and nearby construction is combustible, or if combustibles must be stored in the area, encase tank legs with materials having adequate fire resistance. If this is not done, instructions should require personnel to leave the area when fire threatens tank legs.

IV. Combination with other chemicals. All occupancies involving ap-

Special Chemical Hazards

preciable quantities of acids or alkalis should be carefully analyzed by fire prevention personnel to assure plant safety. This action and the recommendations stemming therefrom should include not only the corrosive chemicals but also other materials that might react with them. A good example of the problem is any large electroplating plant. At such a plant, it is probable that most of the following would be present in substantial amounts: strong acids in glass carboys, solid caustics in drums, miscellaneous flammable solids, cyanides in lined fiberboard drums, chromic acid in pails, and miscellaneous flammable liquids. Activities likely to be present would include plating, metal cleaning, degreasing, rinsing, and descaling.

In this type of occupancy, it is essential that noncompatible items be kept apart both in operations and in storage and that other suitable precautions be taken. For example, while sodium cyanide is a health hazard when improperly handled, it is noncombustible and relatively stable. However, if sodium cyanide comes in contact with acids, hydrogen cyanide is generated. Hydrogen cyanide is a severe health hazard, highly combustible, and chemically unstable. Because of potentials such as this, it is essential that fire prevention personnel recognize that, even though corrosive chemicals may not be important from the standpoint of their own combustibility, they can play a vital part in the overall fire safety picture.

Toxic Chemicals

Many of the products of combustion are toxic, but they differ greatly in degree of toxicity. For example, smoke from burning vegetation is only dangerous in large quantities. In contrast, many commercially used materials are extremely toxic in small quantities. Other commercial products produce very toxic gases under fire conditions. Some of these highly toxic materials are combustible, and others are not. For materials of this type, however, the point of concern is whether or not they would present a hazard to life in the event of fire. For present purposes, toxic materials are defined as those materials that, when inhaled, ingested, or absorbed through the skin—in small quantities—can cause serious injury or death. Certain corrosive chemicals and a few halogens because of their special nature are omitted from this section and will be commented upon elsewhere.

I. Uses. Toxic materials are widely used and, so long as their potential hazard is recognized and guarded against, form an important contributing factor to the American standard of living. Some of the operations where toxic materials are regularly used are listed below.

Other industries could of course be added, but the list is sufficient to indicate how widespread the use of toxic materials is.
  A. Laboratories.
  B. Drug manufacturing and supply houses.
  C. Manufacture, storage, and use of fumigants, pesticides, insecticides, and herbicides.
  D. Heat treating (cyanides, ammonia).
  E. Manufacture, storage, and use of aniline inks and dyes.
  F. Electroplating (cyanide baths).
  G. Certain missile and aerospace manufacturing, storage, and usage operations (organic fluorides, nitric oxides, fluorine, hydrazine, and the like).

II. Major toxic chemical groups. There are literally thousands of toxic chemicals in use. Fortunately, most of these chemicals fall into a few groups, of which the following are the most likely to be encountered.

  A. Alkaloids. This group includes atropine, cocaine, morphine, quinine, strychnine, and similar compounds. All of these chemicals are used for medical purposes, and strychnine is also used for predatory animal control. All can be very poisonous, and vapors from liquid forms are toxic.

  B. Aniline and its derivatives. Aniline is primarily used in dyes and inks. It is usually in the liquid state and can be absorbed through the skin. Vapors from the liquid can damage respiratory organs. Threshold limit for allowable concentration is usually considered to be about 5 ppm.

  C. Antimony compounds. These compounds are used in printing operations involving type metal. Fumes from the molten metal are both toxic and irritating. This is one of the reasons for requiring good ventilation around type-melting operations.

  D. Arsenates and arsenites. One of the principal uses of these compounds is in the manufacture of insecticides and pesticides. Threshold limits are very low.

  E. Aromatic hydrocarbons. In this group are such compounds as benzene, toluol, and xylene. These chemicals are commonly used in fabric-coating operations, particularly rubber spreading. Vapors in any concentration can result in acute poisoning. In addition, the compounds are flammable.

  F. Creosols. At ordinary temperatures, creosols are liquid. Toxicity rating is 5 ppm.

Special Chemical Hazards 201

G. Cyanides. The danger of these compounds is generally recognized and requires no comment here except to note that many of them react with acids or moisture to form hydrogen cyanide gas, which is extremely deadly as well as highly combustible. Carbon dioxide extinguishers do not belong near cyanide baths or storage. Carbon dioxide discharged from the extinguisher would combine with moisture in the air to form carbonic acid. The acid, in turn, would react with the cyanide to form hydrogen cyanide.

H. Cyanogens. All are water soluble and poisonous. Cyanogen proper may be used as either a liquid or a gas, but it is usually found as a gas. In either form, it is poisonous and highly flammable.

I. Fluorides. All are poisonous. Inorganic fluorides are noncombustible. Organic fluorides are combustible.

III. Special hazards.

A. Some poisonous liquids, such as the creosols and anilines, give off highly toxic vapors when exposed to elevated temperatures. Some liquids, such as toluene, can give off toxic vapors if spilled at room temperatures. At this point spelling needs to be emphasized. Benzene is a highly toxic chemical, whereas benzine has a relatively low toxicity rating.

B. Certain toxic gases have very little odor, and some of these are also colorless—for example hydrogen cyanide. Other poisonous gases, such as hydrogen sulfide, are colorless and have a strong odor which, however, quickly paralyzes the sense of smell.

C. Many bathroom bowl cleaners contain a fair amount of hydrochloric acid. Mixing a cleaner of this type with bleaching agents may release substantial amounts of free chlorine. Bulk quantities of these two kinds of materials should be stored in separate areas.

IV. Shipping containers. It is essential to note that shipping cylinders for toxic gases differ sharply from those used for most gases. Cylinders containing certain poison gases, such as hydrogen cyanide, phosgene, and cyanogen, do not have any pressure-relieving devices and may rupture violently when exposed to high temperatures. These cylinders should never be stored where they would be unnecessarily exposed to fire or other causes of unusually high temperatures. Certain other toxic gases, such as sulfur dioxide and chlorine, are shipped in green-label, Department of Transportation–approved cylinders, which do have pressure-relieving devices. These cylinders are not likely to explode, but their contents are toxic, and they should also be located where they are not likely to be exposed to fire. Large ammonia cylinders (green label) have devices to relieve pressures of 165 lb. or more. Most smaller ammonia cylinders do not have pressure relief devices and may rupture during a fire.

Fluorides and cyanides are soluble in water. Their containers sometimes are wax or polyethylene bottles (for hydrogen fluoride) or fiberboard drums (for copper cyanide). Such containers may fail during a fire, or they may be weakened by water applied to a fire and collapse. In either event, there is poisoned water to dispose of. Careful planning would preclude such potential disasters.

V. Fumigants. Fumigants are gases, liquids, or solids from which toxic gases are produced to exterminate harmful organisms. In many instances, the fumigant is mixed with a flammable solvent carrier, and the mixture is sprayed into the room being fumigated. (The same is also true of many insecticides and pesticides.) Most fumigants are also toxic to man, to higher forms of animal life, and to plants.

  A. Kinds of fumigants. Some of the more common fumigants are listed below.
   1. Acrylonitrile liquid.
   2. Aluminum phosphide solid.
   3. Benzene liquid.
   4. Carbon disulfide liquid.
   5. Ethylene oxide gas.
   6. Hydrogen cyanide gas or liquid.
   7. Methyl bromide gas.

  B. Hazards. All of the above fumigants, as well as others in use, are highly toxic. Some are highly flammable, some react violently with metals, some react with water or other substances. Before a fire prevention officer inspects a place where it is known that fumigants are used in quantity, he or she should refer to a reliable hazardous chemicals handbook (for example, the NFPA *Fire Prevention Guide on Hazardous Materials*) to ascertain what the hazards of the particular material are.

  C. Fumigation safety measures. Elementary precautions that should be taken prior to fumigation are listed below.
   1. Extinguish all potential ignition sources.
   2. Clean the area to remove any unnecessary combustibles.
   3. Tightly seal the building or room, and post warning signs.
   4. Provide watchman service.
   5. Notify the fire department.
   6. Maintain appropriate installed protection in service.

Special Chemical Hazards 203

D. Safety measures for accidental release. It is, of course, possible to have accidental release of fumigants. Preplanning for such incidents should include such measures as the following:
  1. Provide for evacuation.
  2. Provide for disposal of contaminated water.
  3. Furnish means to properly ventilate or neutralize toxic gases.
  4. Maintain contact with professional fumigators as necessary.

E. Uses. Areas and operations where the use of fumigants may be anticipated include:
  1. Buildings, boxcars, ships' holds, and vehicles.
  2. Grain and other commodity storage facilities.
  3. Some types of packaging operations.
  4. Fabric fumigation.
  5. Soil fumigation.
  6. Greenhouses.
  7. Various manufacturing, storage, and sales locations.

VI. Etiological agents. Etiological agents are substances that cause certain important diseases. Even though the United States has abandoned the idea of biochemical warfare, there is no assurance that all other nations have done the same. Therefore, it is reasonable to assume that the United States is still conducting civilian research to develop antitoxins, serums, and other means of protection against possible epidemics. Some of the supergerms that are known to have been developed in the past cause anthrax, cholera, diphtheria, rabies, bubonic plague, or yellow fever. Even though any such supergerms may exist in only few locations, the problem is potentially too serious to be taken lightly. The author has inspected several biological research facilities and discussed the problem with top ranking experts in the field. It must be understood that the release of supergerms could result in millions of deaths.

As in everything else, prevention is the first line of defense. Preventive measures against the spread of etiological agents include, but are not limited to, the following:

A. Sealing off buildings housing such operations from other buildings.
B. Fire resistive construction.
C. Metal furniture.
D. Minimum combustibles in the occupancy.
E. Air locks.
F. Decontamination chambers.

G. Tight control of all wiring and piping.

H. Special design for air conditioning, ventilating, and heating systems.

I. A preplanning-for-emergency program that includes necessary advance coordination with fire and police departments.

VII. Pesticides. Strictly speaking, the term pesticides includes all toxic compounds designed to kill pests of all types. As used here, however, the term denotes insecticides, herbicides, and fungicides. These poisons may be used undiluted, they may be diluted in water or oil, or they may be mixed with paint or varnish. These materials have a considerable range of toxicity, but all of them are potentially dangerous to humans if misused.

A. Uses. Like fumigants, pesticides are found in many locations, including stores, agricultural chemical warehouses, farm storage, plant nurseries, transportation vehicles, and manufacturing plants.

B. Hazards. In addition to their toxic effects, pesticides have other potential hazards. Many are combustible and are used in a powder form. If such pesticides are dispersed in the air as a dust cloud and come in contact with an ignition source, they may explode. Likewise, flammable solvents in fogging operations can, given an ignition source, set up a violent vapor explosion, even though the pesticide itself may be noncombustible.

C. Safety measures. Pesticides, at least in appreciable quantities, should never be located where they would be susceptible to serious fire exposure. Container rupture in the event of fire could spread fire throughout a wide area.

If we had 100 percent effective fire prevention—and we do not—bulk pesticide operations would be conducted in fire resistive buildings with a minimum of combustibles present. Unfortunately, such materials are often stored in wooden buildings or in the same area as combustible materials. The combustibles should, if at all possible, be relocated to safer areas. In addition, provision should be made in advance for controlling runoff during a fire, so as to prevent serious problems elsewhere. Parathion, for example, has a toxic threshold toxic rating of only 0.1 mg per m^3, and this is not the deadliest pesticide. It is also regrettable but true that many toxic chemicals, including combustible materials, are stored in moderate quantities in glass bottles and paper containers in stores where ignition sources are present, and the containers are within the reach of small children. Fire prevention personnel should take a closer look at such situations.

Special Chemical Hazards 205

At a minimum, fire prevention preplanning should, as appropriate, provide for the following:
1. Segregation of pesticides from other chemicals and combustible materials.
2. Guardrails for storage shelves holding pesticides in glass bottles.
3. Sprinklered, fire resistive buildings.
5. Special drainage and ventilation to such safe areas as holding tanks. Filters and similar equipment should be noncombustible.
6. Employee training programs.
7. A list of on-call personnel qualified for emergency actions.
8. Warning signs.
9. Emergency medical systems.
10. Evacuation programs.
11. Coordination with fire and police departments.

VIII. Summation. Toxic materials, whether combustible or not, are a serious fire problem, and they have been involved in a number of major fires. There is a considerable body of literature on this subject, but perhaps the best single source is the office of the state fire marshal of California. (This state uses a tremendous variety and quantity of pesticides in agriculture.) Fire prevention personnel who are likely to be involved with toxic chemical fire problems should acquire suitable reference literature from qualified sources such as state fire marshals' offices.

Explosives

The commercial use of explosives is not new; most of these substances have been around for a long time. However, in recent years a new element has been added: rocket propellants. The propellants are not, strictly speaking, explosives, and they are not used as such. However, they have many characteristics in common with explosives and, in a general fire prevention textbook, may well be discussed under that heading.

Accidents with commercial explosives are not frequent because great care is normally used in handling these materials. However, many of the accidental explosions that have occurred have resulted in loss of life and millions of dollars worth of damage to property. Despite the detailed regulations governing the manufacture, storage, transportation, and use of explosives, the potential hazard is so great that one should never assume that there will never be an accident or a failure to live up to basic safety precautions. For this reason, all fire prevention

personnel should understand the basic precautions and have access to applicable regulations for the explosives and operations involved. To be prepared for emergencies and to assist in those actions necessary to prevent emergencies from occurring, it is necessary to know the following:
1. The kinds and amounts of explosives and how they act.
2. What other materials may be stored or handled in the explosives area.
3. Where and how explosives are stored.
4. What people and property may be exposed in the event of an explosion or of a fire endangering the explosives.
5. Whether or not an explosives carrier is accompanied and what its route is. Military explosives carriers normally have escorts. Some states and cities prescribe escorts for civilian explosives cargos, but others do not.
6. What protection is provided for buildings where explosives are manufactured or stored.

I. Explosives groups. There are four groups of explosives: three classes and a forbidden group. The last may not be transported in interstate commerce.

A. Class A. These materials are the detonating type of explosives. This group includes nine subdivisions, which are of interest primarily to the expert.

B. Class B. This group presents a flammable hazard and is not subdivided. Materials function by rapid combustion and not by detonation. Even though detonation is not a problem, this group does include some of the most dangerous explosives.

C. Class C. Explosives in this group contain small amounts of Class A or Class B explosives or both. They present a minimal hazard because of the limited amounts present.

D. Forbidden explosives. This unacceptable group comprises the following materials.
1. Materials subject to spontaneous ignition or pronounced decomposition at temperatures below 160° F.
2. Materials that contain both an ammonium salt and a chlorate.
3. Liquid nitroglycerin.
4. Condemned explosives.
5. Certain special fireworks.
6. Explosives in leaking or damaged packages.
7. Any newly developed explosive until it is classified.

Special Chemical Hazards

II. Markings. All explosives containers should be marked with their classification, including the type number for Class A materials. Other pertinent information to identify the hazard should also be noted on each container.

III. Basic transportation requirements. A number of standard safeguards have been set up for the transportation of explosives, and their use must be mandatory if major disasters are to be prevented. Among the most important of these precautions are the following:

A. All personnel should be thoroughly trained with respect to the explosives for which they are responsible and in emergency actions to be taken in event of fire or accident. For example, high explosives subject to a fire or heat from a fire may melt, flow, drip, or spread and mix with the ground and wreckage. In this condition, they are extremely hazardous, and the area should be secured pending disposal operations by demolition experts.

B. No vehicle, train, ship, or aircraft carrying explosives should be left unattended at any time.

C. Explosives and fuses or detonating caps should never be carried in the same vehicle, railway car, ship's hold, or aircraft.

D. Transportation equipment must be kept in a high state of repair and frequently inspected. For trucks, inspections must include such equipment as brakes, lights, mufflers, fuel lines, ignition systems, and tires.

E. Smoking near explosives must be prohibited.

F. Refueling must be done with careful attention to safety and only when absolutely necessary. Whenever possible, all fueling should be accomplished prior to or after unloading of explosives.

G. Parking of loaded explosives trucks in populated areas, even with attendance, should be prohibited. Similar precautions should be taken with other forms of transportation.

H. Do not transport explosives in trailers.

I. On trains, explosives cars must not be located next to the locomotive, the end of the train, or hazardous cargo. Explosives cars must also be separated from each other. Preferably, the separating cars should be noncombustible and either be empty or contain a noncombustible cargo.

J. Ships must not be loaded with explosives at a common carrier pier, and explosives ships must not be blocked at a pier.

K. Never store, garage, or repair loaded trucks in buildings. Remove the explosives first.

L. Never allow a train with explosives cars to be placed in a "hump" yard. When explosives are involved, car separation is a ticklish proposition.
M. Install danger signs on all carriers.
N. Make sure explosives containers are properly blocked, braced, and secured before getting underway.
O. Train crews should be particularly alert to hot bearings and hot wheels. (These conditions are caused by hard, continuous braking on hills.)
P. When explosives are involved in a fire, evacuate personnel to a safe area.

IV. Some transportation fires.
A. A few years ago, an "ammo" train blew up in a Roseville, California, freight yard. More than 300 people were injured, and property damage was estimated at $35,000,000.
B. In 1967, burning JP5 fuel enveloped a plane and its bomb load on the carrier *Forrestal*. One hundred and thirty-one men died, and the ship was heavily damaged.
C. In 1959, a truck blew up in Roseburg, Oregon. Twenty buildings were destroyed, and ninety more were heavily damaged.

Fires followed by explosions or explosions followed by fires have almost always resulted in severe property damage and frequently have resulted in numerous fatalities and/ or injuries.

V. Storage safeguards. Many of the precautions cited for transportation vehicles are also applicable to storage locations. In addition to the common safeguards, the following precautions should be adhered to.
A. Each room or building containing explosives should be posted as to the number of personnel permitted in the area and the type and amount of explosives that are permissible. These limits should be rigidly enforced.
B. Protect explosives against excessive heat (keep them under 90° F. if possible) and cold.
C. Protect against dampness, lightning, and flying objects.
D. Store separate classes of explosives in separate magazines.
E. Handle packages gently and do not open or repack a package inside a magazine.
F. Only wooden tools should be used to open and close packages.
G. Housekeeping must be emphasized.
H. Only electric flashlights or lanterns should be used for portable lighting.
I. Never allow loose dynamite or powder to be exposed.

Special Chemical Hazards

J. Prohibit smoking. A space outside in a safe area should be set aside for the deposit of matches and lighters. Spot searches for matches are also desirable.

K. Doors to magazines should be kept locked, and adequate watch service should be provided.

L. Make certain manufacturers' recommended special safeguards are carried out.

M. Preferably, permanent storage buildings should be fire resistive or noncombustible and located in well-drained areas that are free from the possibility of exposure fires.

N. Buildings should not be located on rock strata that can carry shock waves to important buildings.

VI. Storage fires. To illustrate the potential for disaster where appreciable quantities of explosives are stored, it might be well to comment on the Richmond, Indiana, sporting goods store fire in 1968. Approximately 1 ton of smokeless and black powder stored in the basement exploded. As a result of the explosion and the ensuing fire, 41 persons died, more than 100 people were injured, and some fifteen buildings were destroyed.

VII. Special blasting precautions.

A. Keep igniters separate until they are used.

B. Personnel should be alert to the possibility of stray electric currents, static electricity, and radio frequency currents.

C. Where possible, power circuits in the blasting area should be grounded.

D. To eliminate induced current hazards, blasting wires should be laid at right angles to power circuits.

E. Blasting wires should be separated as much as possible from static-generating equipment.

VIII. Manufacturing processes. The many different types of manufacturing processes are too numerous to cover in a single text. Each process should be reviewed separately and full advantage taken of the special literature that is available on these processes.

IX. Rocket propellants. Rocket propulsion systems are designed to propel manned and unmanned spacecraft, missiles, and miscellaneous special purpose vehicles. Systems range in size from units weighing less than a pound to systems capable of moving vehicles containing several people together with supplies and equipment for several months. In the not-too-distant future, there undoubtedly will be intercontinental passenger and freight rockets. The technology to accomplish this advance is here now.

While it is true that currently there are only a few launching pads, it must not be overlooked that the fuels are manufactured elsewhere, stored in many locations, and transported by land, sea, and air. It is with manufacture, storage, and transportation that the average fire prevention engineer, inspector, or trainee should be primarily concerned.

As currently used, rocket propellant systems comprise two basic forms—those using solid fuel and those designed for liquid fuel. Because of basic differences, these systems must be considered separately.

A. Solid propellants. The smaller of the propellant systems uses such solid fuels as nitroglycerine combined with nitrocellulose or ammonium perchlorate combined with polyurethane. Most of the combinations also include stabilizers to prevent premature ignition. In general, solid propellants are used by the military, and their exact composition is classified. However, solid propellants have some features that are not classified, and fire prevention personnel should have background information on these features.

1. Once a solid propellant starts burning, it is virtually impossible to extinguish the fire, and attention should be focused on fires that start in surrounding areas. For this reason, solid propellants should always be shipped separately from the igniters.

2. In the manufacturing process, it is often necessary to trim blocks of propellant to fit rocket motors. This trimming process at times requires the use of a flammable coolant. The author has seen acetone so used.

3. Always consider solid propellants as Class A explosives.

4. In the manufacturing process, critical operations should be protected by ultra-high-speed deluge systems with discharge devices aimed directly at the propellant. Equipment is available that can sense an incipient flame and operate a special quick-release valve that discharges water prior to actual ignition. If the system functions properly, the incipient fire will be doused in milliseconds. If it fails, there will be an explosion. Plants with these systems should have detailed plans and a mandatory program for testing and maintenance by *qualified* personnel. This is no place for amateurs.

5. Equipment handling solid propellants should be operated, tested, and maintained exactly as preplanned by experts. There should be no exceptions and no excuses.

B. Liquid propellants. Most liquid propellants are in the civilian area, and their composition can be pinned down. Some of the com-

Special Chemical Hazards 211

ponents can be preloaded into rockets at any time. Others, such as cryogenic liquids, which must be kept refrigerated until just before launching, cannot be preloaded.

Properly handled, liquid propellants present far less of an explosion problem than do solid propellants. In manufacturing, storage, and shipping, it is standard procedure to keep fuels and oxidizers apart, and this cannot be done with solid propellants. Manufacturing, storage, and transportation of liquid propellants should in general follow the procedures required for flammable liquids and oxidizers. In some instances, however, special precautions are necessary. At an actual launch or a test firing (which may be done at a plant), the components have to be brought together. To do this safely requires an exact timetable for loading.

C. General. There is no practical way to prescribe specific rules for handling rocket propellants as a class; there are too many variables involved. If an inspection is to be made of a propellant facility, the only safe procedure is observe the following guidelines.

1. Determine what chemicals are involved. Do not be disturbed by unpronounceable chemical names, which are chiefly of interest to the chemist. Virtually all propellants fall into one of several groups that have relatively simple names. With a few exceptions, each of these groups has fairly common fire characteristics, and these characteristics form the point of interest.

2. Investigate the behavior characteristics of the hazardous items, and from that develop safe procedures for handling storage, manufacture, transportation, and other procedures.

3. Make certain plants comply with the safety rules laid down by experts. This means watching out for departures from standard operating procedures—for example, bypassing a process or providing inadequate maintenance.

4. Make sure that plants are prepared for such emergency actions as evacuation and that there is coordination with the fire and police departments.

5. Don't guess, know. Don't hesitate to ask a professional in the field for advice. Fireballs and poison gas clouds are not toys.

6. Make certain responding fire departments have had training in handling such emergency situations.

D. Summary. Explosives and rocket propellants, like other hazardous materials, should be neither feared nor ignored; they should be understood. No material, regardless of its properties, need be feared if it is handled with intelligence. This means using our own positive knowledge and that of others to the end that operations are safe.

Cryogenics

Cryogenic liquids are liquefied gases at very low temperatures. For our purposes, cryogenic liquids are defined as those gases that become liquid at temperatures of $-100°$ F. or lower with pressures fairly close to normal at the point that the gas liquefies. This does not mean that these liquids cannot be stored at considerably above normal pressures. They can and frequently are.

I. Uses. The use of cryogenic liquids in industry and in a few other areas is increasing rapidly, and it should be anticipated that the rate of increase in usage will continue to speed up in the future. The reason for this is that cryogenic liquids perform very well in automated mass production and, in many instances, require less handling than do compressed gases. At present, the principal areas of use are those listed below.

A. Aerospace industry (oxygen, hydrogen, helium).
B. Metal fabrication (oxygen).
C. Cold storage and food freezing (several gases).
D. Hospitals (oxygen).
E. Gas companies (methane).
F. Other industries with large gas piping systems for oxygen, hydrogen, nitrogen, or helium.

Cryogenic liquids in smaller quantities are used in many industries.

II. Gas laws. Before attempting to analyze the fire potential of any existing or proposed use or storage of cryogenic liquids, it is essential that there be a clear understanding of basic gas laws. As stated previously, these laws can be combined into the following simple formula: $T_1/T_2 = P_1V_1/P_2V_2$. T equals the absolute temperature, P equals the absolute pressure, and V equals volume.

Absolute temperature is based on absolute zero, which on the Fahrenheit scale is approximately $-459°$ F. Therefore, a reading of $-359°$ F. equals $100°$ absolute, and one of $100°$ F. equals $559°$ absolute.

Absolute pressure is recorded pressure plus normal atmospheric pressure (14.7 psi at sea level).

To illustrate the formula and the definitions, assume that a cryogenic liquid at $-400°$ F. has its temperature raised to $100°$ F. Assume that the original gauge pressure is 200 psi and that the volume remains

Special Chemical Hazards

constant. What does the temperature change do to the pressure? Values for the equation are obtained as follows:

$T_1 = 459° - 400° = 59°$. $P_1 = 200 + 14.7 = 214.7$ psi
$T_2 = 459° + 100° = 559°$. $V_1 = V_2$ (cancel out).

By substituting these values for symbols in the formula, we obtain the following results.

$59/559 = 214.7 \; P_2$.
$P_2 = 2,033$ psi.

Unless the container is unusually strong, something has to give. For this reason, relief valves and/or frangible discs are provided.

Taking another tack, assume the same temperature conditions, the same original pressure, and an initial volume of 3 cu. ft. How much space will be filled if the container ruptures and the gas expands to normal pressure?

$T_1 = 59°$. $P_1 = 214.7$ psi.
$T_2 = 559°$. $P_2 = 14.7$ psi.
$59°/559° = (214.7 \times 3)/14.7 \; V_2$. $V_1 = 3$ cu. ft.
$V_2 = 415$ cu. ft.

This expansion ratio should provide clear evidence of the need to provide adequate venting facilities when appreciable quantities of cryogenics are stored in buildings.

III. Expansion. Although such incidents fortunately are rare, a major tank rupture or a transportation vehicle accident sometimes releases a large quantity of a cryogenic liquid. As normally used in industry or carried in a vehicle, cryogenic liquids have expansion ratios ranging from 400 to 1,000. A 1-cu. ft. container of hydrogen with an expansion ratio of 1,000 would, in the event of container rupture, release enough vapor to fill a 10 ft. x 10 ft. x 10 ft. room with pure hydrogen. Since hydrogen has an explosive range of 4 percent to 75 percent, it is theoretically possible for this 1 cu. ft. container to place a space measuring 25 ft. x 100 ft. x 10 ft. in the explosive range. Because of pressure relief valves, proper construction, adequate venting, control of ignition sources, and control of other occupancies, a problem of this type rarely occurs in industry. The consequences of neglect, however, are potentially so disastrous that continuing education is needed.

Even though industrial plants have been relatively free from major cryogenic spills, there have been a number of vehicle accidents that resulted in large spills of 12,000 gallons or more. Such spills can easily form a fireball of huge dimensions and threaten a large area. Accidents of this nature have necessitated halting rail and vehicular traffic for several hours, evacuating large areas, extinguishing ignition sources, and other major emergency actions.

Safety precautions for vehicles transporting cryogenic liquids are prescribed by DOT, do not come within the purview of most fire prevention personnel, and are therefore omitted.

IV. Vapor density. At ambient temperatures, gases have known vapor densities. The choice of appropriate fire safety measures depends on whether the gases in question are lighter or heavier than air. In the cryogenic state, the vapors of all gases, including hydrogen, are heavier than air. Such vapors hug the ground and flow towards low points until the vapors warm up. It is vitally important to remember this, for some normally lighter-than-air gases such as hydrogen are odorless, colorless, and tasteless.

Like other cryogenics, liquefied hydrogen is carried in tank cars of up to 35,000 gallons capacity. If a liquefied hydrogen spill of this size should occur, several hours might elapse before the hydrogen warmed up enough to start dissipating upwards.

V. Cryogenic ratings. Like other materials, cryogenic liquids are rated for combustibility, health hazard, and reactivity. Ratings are 0 to 4, with 0 representing no or negligible hazard and 4 representing the most severe hazard. Table 7 provides ratings for nine cryogenic liquids, including all of those in widespread use.

Table 7

Item	Flammability	Health Hazard	Reactivity	Explosive Range
Ethylene	4	1	2	2.7-36%
Fluorine	0	4	3	
Helium	Inert			
Hydrogen	4	0	0	4-75%
Krypton	Inert			
Methane	4	1	0	5-15%
Nitrogen	Inert			
Oxygen	0	3*	0	
Xenon	Inert			

*Gaseous oxygen from liquid oxygen is readily absorbed by most clothing, and a flash fire from any source of ignition is possible for some time after exposure.

Special Chemical Hazards

VI. Safety precautions. There are two types of responsibility for fire safety of cryogenic liquids: that of fire prevention personnel and that of firefighting forces. This book discusses actions to be taken by fire prevention personnel but omits discussion of the responsibilities of firefighters. Fundamental safeguards for manufacture, tranportation and storage of cryogenic liquids include, but are not limited to, the following:

A. Liquid oxygen.
 1. Make sure manufacturing and storage buildings are fire resistive or noncombustible.
 2. Place oil separators in the air line ahead of the driers.
 3. Provide means for prompt removal of hydrocarbons from rectification towers.
 4. Keep liquid oxygen a safe distance away from above-ground flammable liquids, LPG, gas holders, combustible buildings, and combustible yard storage.
 5. Keep outdoor storage areas free of vegetation.
 6. Do not allow crawl spaces under buildings where debris can accumulate.
 7. Locate outdoor storage on concrete.
 8. Keep ignition sources away.
 9. In piping systems, use only specifically approved equipment, and never locate piping in the same trench with piping for flammable gases or liquids.
 10. Set up prescribed routing for vehicles moving liquid oxygen from plant to destination. Vehicles should stay away from important buildings and on concrete pavement to the extent possible. Liquid oxygen can ignite asphalt paving. All it takes is a little shock.

B. Fluorine. This pale yellow gas has a sharp odor. It is dangerously reactive with oxidizable materials, water, many acids, and certain other substances, and it is extremely toxic. In storage, fluorine must be kept in special containers and away from any of the numerous materials with which it would violently react. Fire prevention personnel should be concerned with the following measures to minimize personal injury and property damage from a major spill or fire.
 1. Capability for safely evacuating people from an affected area.
 2. Means for obtaining prompt assistance from firefighters and police.

3. Provision of protective clothing for those who must remain in an affected area.
4. Elimination of ignition sources.
5. Capabilities for stopping leaks.
6. Training programs for plant fire brigades.

C. Hydrogen. Precautions for handling liquefied hydrogen include, but are not limited to, the following:

1. Containers of liquid hydrogen must vent outdoors and away from personnel and important buildings and equipment.
2. Piping systems should be designed and the installations approved by qualified experts.
3. Areas containing hydrogen control equipment should be vented at ceiling level.
4. Heating systems for liquid hydrogen vaporizers should be of the indirect type only. Such systems include those using steam, hot water, or hot air.
5. Electrical equipment located near controls or in areas where connections are regularly made and broken must be explosion-proof and of the appropriate type.
6. If liquid hydrogen is stored near oxygen or flammable liquids, the hydrogen should be on higher ground than the other substances.
7. Outside storage should be fenced and posted.
8. Under no circumstances should liquid hydrogen be stored anywhere near liquid or gaseous chlorine.
9. Keep all liquid hydrogen storage and bulk use away from concentrations of people.

VII. Summation. This section has discussed cryogenic liquids from the standpoint of large-scale fire hazard potentials, with specific attention being paid to selected liquids. Space does not permit a detailed survey of small-scale operations or of the less commonly used liquids. For these situations, reference should be made to published standards of the NFPA and to the publications of other recognized authorities. These references should be available to fire prevention personnel as required by conditions.

Unstable Chemicals

Many manufacturing operations involve at some stage the use of chemicals which, for want of a better term, are classed as unstable chemicals. Excluding air- and water-reactive substances and true explosives, unstable chemicals may properly be defined as those materials

Special Chemical Hazards 217

that in their pure state or as normally commercially produced are potentially explosive or can polymerize violently, decompose, or undergo other violent changes.

I. Stabilization methods. Because of the potential for violent reactions, it is vitally important to remember that almost all of these unstable materials can be made stable. Virtually all of them fall into a few major groupings, and the appropriate actions to be taken to stabilize materials in each grouping are given in many industrial chemical handbooks. The principal stabilization methods are listed below.

  A. Dissolving in a solvent.
  B. Dampening with water.
  C. Adding an inhibitor. (An inhibitor is the reverse of a catalyst.)
  D. Storing in dark bottles (generally brown or green).
  E. Storing in dilute concentrations.
  F. Keeping material dry.
  G. Avoiding exposure to sunlight or ultraviolet radiation.
  H. Using airtight containers.
  I. Inerting the air space around the material.
  J. Maintaining constant pressures and/or temperatures.
  K. Keeping the unstable chemicals free from foreign material.

II. Uses. Even if a person could remember all the details about unstable chemicals, the knowledge would not do him much good unless he also knew where and how these substances are used. A fire prevention officer should, however, understand the basic problem associated with unstable chemicals, know where these materials may be found, and have available sources of detailed information as needed. Some of the most common unstable materials and their areas of use are listed below.

  A. Acetylene. Cutting and welding; metal fabrication; manufacturing of acrylonitrile, vinyl chloride, and other substances. (Acetylene is fairly stable as used in welding because it is dissolved in acetone and the cylinder has a special filler.)
  B. Acids. Some acids, including acrylic, paracetic, perchloric, and picric acids, are unstable. They are used in food processing, bleacheries, plastics manufacturing, medicine, metallurgy, and elsewhere.
  C. Aldehydes. Textile industry, drug manufacturing.
  D. Ammonia compounds. Explosives, fireworks, propellants.
  E. AZO compounds. Blasting agents.
  F. Butyl compounds. Synthetic rubber.
  G. Chlorine compounds. Food processing, bleaching.

H. Ethers. Anesthetics, explosives, solvents.
I. Fulminates. Detonators.
J. Nitro-compounds. Celluloid, dyes, explosives, heart stimulants.
K. Per-acids. Bleaching, food processing, fungicides.
L. Peroxides. Explosives, medications, plastics.
M. Vinyl compounds. Propellants, plastics.

This list is not all-inclusive, but it illustrates the point that unstable chemicals may be found in almost any industry. The amounts are usually small, but the substances are used at some stage of many industrial processes. Fire prevention personnel should determine whether or not they are used and, if so, where and in what quantities. It is not necessary to do extensive probing for this information; a basic knowledge of groupings and a casual reading of labels in special storage areas or in laboratories is enough to locate the chemicals. In addition, most containers of unstable chemicals carry warning labels. When such chemicals are found, a fire prevention officer should determine whether or not the plant is aware of the hazard and is handling the materials safely.

III. Some types of potential violent reactions and materials that might be involved.

A. Detonations. Acetylene, ammonium nitrate, carbon disulfide, perchloric acid, some organic peroxides.

B. Deflagration. Pyroxylin, organic peroxides.

C. Decomposition (violent). Pyroxylin, certain peroxides, methyl parathion.

D. Formation of explosive peroxide mixtures from various ethers.

E. Polymerization (violent). Acrylonitrile, butadiene, ethylene oxide, vinyl chloride.

F. Miscellaneous. Many unstable chemicals may break down to form lethally toxic materials. Virtually all unstable chemicals are capable of forming breakdown products with serious injury potentials.

IV. Typical safety precautions. Unstable chemicals do not get in the news very often because most users and most transportation concerns are well aware of the hazards and exercise careful supervision. Nevertheless, accidents do happen, and for these we should be prepared. A little practical knowledge always helps. Some examples of precautions that can be taken include, but are not limited to, the following:

A. A few unstable chemicals are self-reactive and can explode with very little encouragement. These materials should be stored and used in the smallest possible amounts.

Special Chemical Hazards 219

B. Fluorine is capable of reacting with almost anything. It should therefore be kept in the original containers until used.

C. As noted earlier, it is possible to stabilize many unstable chemicals. Plants and storage locations should have personnel trained in stabilizing operations as well as the materials necessary for stabilizing.

D. Acetylene or similar gases should never be present where oxygen, particularly liquid oxygen, is being manufactured. Traces of acetylene entering an air intake have combined with metallic elements in certain portions of the oxygen piping to form an extremely sensitive explosive.

E. Pyroxylin, for all practical purposes, is no longer used in nitrate film, but it is used in other areas. Particularly sensitive are those where pyroxylin lacquer is used or where some types of drafting tools and training aids are made. In addition to observing the precautions cited in NFPA standards, it is important to check fire protection. Cellulose nitrate carries its own oxygen, and a fire in this compound requires cooling for extinguishment.

F. Ethers present several kinds of hazards.

1. Some ethers are capable of spontaneously forming explosive peroxides in storage. Inhibitors to prevent this from occurring should be on hand.

2. If an ether bottle contains appreciable solids at the bottom, it should be carefully removed from the premises for disposal. If there are any questions as to how to do this, the problem should be turned over to an EOD squad for action.

3. Some ethers tend to cling to tank walls when a tank is emptied for cleaning. There are methods for removing residual vapors from tank walls, and advice on this can be obtained from military ordnance detachments.

G. Perhaps the most common error in dealing with peroxides is lack of attention to relative strengths. For example, 3 percent hydrogen peroxide is a mild medical aid, but 91 percent hydrogen peroxide is capable of exploding from shock, fire, or friction. Most peroxides in use are so diluted that they pose no great problem. There are, however, enough instances of undiluted or slightly diluted peroxides to warrant close attention to relative strengths.

H. When unstable chemicals have been involved in a spill, leak, or accident, they must be moved to a safe area as soon as possible and there be neutralized or otherwise rendered harmless. *Under most conditions* attempts to salvage unstable chemicals are both futile and *foolhardy*. Inert materials, large volumes of water, soda ash, or other protective agents should be available as needed for neutralizing purposes in event of a spill.

I. Plants and warehouses handling unstable chemicals should have continuous training programs for designated personnel and close coordination with responding fire departments.

J. Unless there is positive information to the contrary, for the protection of personnel it should be assumed that unstable chemicals may explode and that they are capable of producing toxic or corrosive vapors or particles.

V. Fire experience. Fires in unstable chemicals are not as frequent as some other types of fires, but these substances have been involved in a number of major fire and/or explosion incidents in which there were multiple casualties and losses in the millions of dollars. Early in the author's career, he experienced an example of the hazards from such chemicals. While waiting for a train, he witnessed an acetylene explosion at a plant from $\frac{1}{2}$ to $\frac{2}{3}$ mile away. In minutes, the plant was virtually levelled, and several employees were killed. Pieces of debris fell within a few hundred feet of the station platform. This was the author's introduction to the unstable chemicals problem.

VI. General. There is no stock solution to the unstable chemicals problem; there are too many variables. What is needed is a commonsense approach, starting with the ability to recognize the hazard. Depending on the extent of the inspector's knowledge of the hazard in question, he or she should not hesitate to seek information from references that are available or from experts on the substance in question. No one can afford to guess with these materials.

chapter 11
Arson Detection and Investigation

To gain a realistic understanding of how fire prevention organizations and personnel can and should play a major role in efforts to reduce the toll in lives and property from arson fires, it is necessary to know five kinds of information: (1) the size of the arson problem, (2) what motivates arsonists, (3) basic arson law, (4) how to make an arson investigation, and (5) how to preserve and present evidence so that it will stand up in court.

Whenever the above five items are understood and put into practice, there is an immediate reduction in the number of arson fires and in the amount of arson losses. All five items, including the legal ones, are well within the capabilities of fire organizations and personnel to learn and/or carry out. Even though few fire safety personnel have any legal background, they do or should have the technical knowledge necessary to provide the assistance that trained legal officers need to successfully prosecute a case. The required technical know-how includes the ability to (1) detect signs of arson, (2) collect and preserve evidence, (3) maintain legally acceptable records, and (4) be an effective witness.

The Problem

I. Losses.
   A. Direct losses. Since 1950, the number of arson fires and the losses in lives and property due to them has risen sharply, as shown in Table 8.

Table 8
Incendiary and Suspicious Fires

Year	No. Fires	Dollar Loss	Source
1950	5,600	$ 15,100,000	Published NFPA Figures
1968	49,900	$131,100,000	Published NFPA Figures
1971	72,100	$232,947.000	Published NFPA Figures
1974	114,000	$550,000,000	Published NFPA Figures

Notes: 1. Statistics include all types of set fires. However, the bulk of the fires are unquestionably arson. 2. Loss statistics do not include arson fires in vehicles, ships, forests, or grasslands.

Along with the increase in arson fires, there has been a substantial increase in the number of fires for which the cause is listed as unknown. Currently, it is estimated that the number of fires in this category number nearly 200,000 annually and that the dollar losses therefrom total approximately $1 billion a year.

B. Business losses. The published loss statistics, horrendous as they are, do not include the major business losses that are incurred any time a business is destroyed or seriously damaged by fire. These losses can and often do exceed direct property losses. For example:

1. Fire destroys the local supply source for plant X. Supplies have to be obtained from a distant source at double the price.

2. Fire destroys plant Y, which manufactures parts that are used in assembly plants within a radius of 1,500 miles. The assembly plants, although not involved in the fire, are forced to shut down or curtail operations because of the fire.

C. Losses from inflation. A factor that is often ignored in computing the cost of any fire is inflation. Today's economy is in an inflationary cycle that at varying rates will probably continue for years. Almost invariably, therefore, rebuilding and replacement costs substantially exceed original building and procurement costs.

To make matters worse, the climbing arson rate actually contributes to inflation. It is a fact that arson fires are increasing at a rate that is far out of line with population increases. It is also true that dollar losses from arson fires are increasing far more rapidly than the rate at which productivity is increasing.

D. Other losses. There is no way to put a price tag on deaths and injuries, nor should any attempt be made to do so. It is possible, however, to put a dollar value on lost earning capabilities, medical costs, increased welfare costs, and certain other indirect losses. The cost of such losses undoubtedly totals more than $1 billion annually.

Arson Detection and Investigation

II. Key factors in the arson trend. There have been a number of investigations into the rising arson trend, and all of them have listed three key factors as largely responsible for the situation. These factors are as follows:
   A. The breakdown in social discipline.
   B. Poor investigative work by arson and fire investigators.
   C. Slap-on-the-wrist punishment.

Of the three, the judgment on the caliber of inspections should be of primary concern to fire prevention personnel. The reports make clear that the reason for the poor results is not carelessness or a lack of technical competence. It is rather a lack of training in how to handle an arson investigation. This missing tool should be provided.

III. Where arson occurs. Part of the solution to the arson problem is to know where the bulk of the arson fires occur and why certain categories of buildings are vulnerable to arson. Arson fires do occur in every building category, but the great majority of them occur in the eight following categories.
   A. Detention facilities.
   B. Hotels.
   C. Churches.
   D. Schools.
   E. Apartment houses.
   F. Department stores.
   G. Dormitories.
   H. Supermarkets.

When this list is analyzed, it should be noted that no basic industrial class is represented. This is no accident. Most industrial plants and large commercial enterprises are protected around the clock by sprinkler systems, watchman service, armed guards, supervised intrusion alarms, fencing and floodlighting, or other safeguards as appropriate. These forms of protection are for the most part lacking in areas where arson is far too frequent an occurrence. This does not mean that arson is not a serious problem to industry—it is. But arson is not as frequent there as for generally unprotected property.

IV. Patterns of arson. A second step towards finding a solution to the arson problem is to realize that, while motives for arson differ widely, within any one group there are recognizable patterns. Knowledge of these patterns is very helpful in coming up with proper corrective measures. Two examples of arson patterns follow.
   A. Loan shark fires. Any area with a loan shark setup is susceptible to arson fires. It doesn't matter whether the area is the waterfront,

the ghetto, a depressed area, or another kind of area. The sequence of events is always about the same. A client in a low or medium-income bracket takes out a loan at a high rate of interest. He insures his property generously, and the insurance is payable to the loan company. The client is unable to keep up payments. His property suffers a total burnout and is sold to the insurance company.

As a result of such events, insurance companies pay out large sums to convict arsonists. There are, in fact, arson-for-profit rings. These rings may include people or firms in the grip of loan sharks, "torches," the loan sharks themselves, and occasionally public adjusters.

B. School fires. In school fires, fraud is rarely a factor. Two thirds of all school fires are set by intruders, not for personal gain or for revenge against a particular individual, but for such prosaic reasons as dislike for the educational process and real or fancied grievances against the system. Schools are in general easy targets. School fires can, therefore, largely be eliminated by providing better security against breaking and entering, by providing suitable detection and extinguishing systems, and through similar protective measures.

V. False ideas about arson. A third step in solving the arson problem is to switch from a negative to a positive attitude by scrapping five false ideas on the subject. These ideas are held by far too many people, including many in the fire safety field.

    A. Arson is the hardest crime to detect. (It isn't.)
    B. You almost have to see them light the match. (You don't.)
    C. The evidence is always destroyed by the fire. (It isn't.)
    D. It was too far gone to tell where the fire started. (It wasn't.)
    E. Arson investigation is not a fire prevention responsibility. (It is.)

VI. Summation. Arson is a serious and growing problem, but present trends can be reversed. What is required is a combined effort by all fire service organizations, including fire prevention organizations, to learn how to detect signs of arson, to investigate, and to follow through so that more cases are solved and more convictions obtained.

About the first thing an arson investigator should learn is what motivates an arsonist.

Motivations for Arson

I. Definition of terms. In any study of arson motivation, it is essential that certain commonly used terms be clearly defined at the beginning.

A. Incendiarism is the deliberate setting of fire without intent to control, regardless of the underlying motive. This term includes both arson and the setting of fires by children or mentally irresponsible persons.

B. Vandalism, from the fire standpoint, is the malicious burning of property for destructive purposes without any motive for personal profit or vengeance. In some states, vandalism by fire is governed by arson law. In other states, vandalism by fire is covered by separate vandalism laws.

C. Basic arson is the setting of fire for personal profit, for personal vengeance, or to exert psychological pressure.

D. Pyromaniacs are mentally unbalanced persons who set fires to see things burn but who have no malicious intent.

E. Espionage incendiary fires are not arson. These fires are set by enemy agents for foreign political or military interests. Such fires are covered by espionage acts and do not come under arson laws.

F. Children below the age of reason may set or cause incendiary fires, but they cannot be guilty of arson.

In each type of case, it is essential that the investigation be thoroughly and properly oriented. The investigating procedures for each type are much the same, but the language used must be chosen with care. The function of the fire prevention inspector is to develop facts and not to make charges. In general, it is unwise for investigators other than fire marshals with legal authority to use the term arson in a fire report. It is far better to use the term incendiary origin or words to that effect.

II. Why people set fires.

A. Some small children and senile oldsters get a childish pleasure out of watching flames. They have no real idea of the harm they may be doing. Control of such incendiaries is simple.

1. Keep matches, lighters, flammable liquids, and similar equipment out of reach.

2. Provide the level of attendance and supervision by others that is necessary.

B. Pyromaniacs get pleasure out of watching things burn and are attracted to major fires. These people are potentially very dangerous and in need of psychiatric care rather than prison. (No firebug was ever reformed by prison.) Pyromaniacs, however, can be easily recognized by their mannerisms and repeated presence at large fires.

C. Vandals set fires—and wantonly destroy property by other

means as well—for the sheer love of destruction. They follow no set patterns of incendiarism, and their apprehension is almost entirely a police function.
D. Some arsonists work for profit.
E. Some arsonists set fires for vengeance.
F. Espionage agents are normally active only where national defense is involved. Fires in critical industrial plants, if arson is suspected, should be investigated by counterespionage agencies.

III. Arsonists' motivations. Fires in the two arson categories above are set for a variety of reasons.
  A. Arson for profit.
    1. Profit to owner.
      a. Liquidating an unwanted or unprofitable business enterprise.
      b. Settling an estate.
      c. Collecting for goods for which there is no market.
      d. Terminating an unprofitable partnership.
      e. Removing buildings that are no longer useful.
      f. Averting a business failure.
      g. Arranging for the sale of land without the cost of removing buildings the new owner does not want.
      h. Saving the cost of demolishing condemned buildings.
      i. Collecting on overinsurance.
      j. Cutting the cost of relocating to another site.
      k. Reducing or eliminating overstocked or obsolete inventories.
      l. Covering the inability to fulfill contracts.
      m. Getting out of business quarters that are too small.
      n. Getting coverage for a cash shortage.
      o. Covering loss of customers to more aggressive competitors.
      p. Saving the cost of upgrading facilities to meet current standards.
      q. Obtaining unjustified tax write-offs.
      r. Destroying competition.
      s. Compensating for cutoff of raw materials.
      t. Saving the operation from being tied up in court by environmentalist suits.
      u. Eliminating energy shortage problems.
    2. Profit to other than owner.
      a. Adjusters—to secure contracts to adjust losses. Adjusters may be in collusion with owners and others to split the profits.

b. Insurance agents—to stimulate business.

c. Building contractors—to obtain contracts for rebuilding or demolishing damaged property. This type of contractor also is not above burning a building to save demolition costs.

d. Workmen—to secure employment as guards, policemen, firemen, or the like.

e. Salvagers—to get contracts for handling the purchase or salvage of damaged goods and other salvaged materials.

f. Criminals—to cover a crime, particularly robbery.

g. Owners and others—to stifle competition or to obtain preferential treatment.

B. Arson for personal vengeance or to further a cause.

1. Labor disputes (sabotage, strikes, lockouts, union versus non-union struggles).
2. Riots.
3. Hatred on a personal level (revenge, spite, jealousy, feuds).
4. Criminal.
   a. To cover murder, destroy records, destroy evidence.
   b. To divert attention in order to burglarize the premises.
   c. To break out of confinement.
   d. Drug addiction.

Although there are other motives for arson, those cited above account for more than 90 percent of all arson fires. If care is taken in the investigation of arson fires and in the reading of arson reports by others, it will become evident that, for each type of motive, there is essentially a standard pattern followed by the arsonist. Once this background has been accumulated, it becomes much easier to nail down solid facts for use by the prosecuting attorney.

Arson Law

The primary impact of arson law on fire prevention personnel is in how they carry out their responsibilities. Fire prevention personnel are not lawyers and have no direct role in determining the degree of criminality involved in any known or suspected arson case. They should not, by word or act, imply that crime was involved in a fire; this is for the prosecutor to prove or charge. For example, it is in order to state in a report that a fire appeared to be incendiary in origin, that fellow employees stated that they saw John Doe set the fire, and to provide other data. It is not in order to say the fire appeared to involve arson. To so state implies criminal acts, and such statements do not belong in a technical report. However, in suspected arson cases, fire prevention personnel have the responsibility of providing legal authorities with information that is as complete and accurate as possible and in a usable form.

I. Laws affecting fire prevention personnel. Most arson laws are concerned with the legal authority granted to such officials as fire marshals, fire commissioners, and fire chiefs and their authorized representatives. Such laws do not directly affect fire prevention personnel. There are, however, certain laws that directly affect the fire prevention inspector.

A. Under the laws of most states, the officer in charge at a fire scene is authorized to take and preserve any property that indicates the fire was intentionally set. In general, such seized evidence is placed in the custody of the fire prevention organization. The owner may petition the court for the return of his property. If the court so orders, the property is returned subject to court orders for the preservation of evidence. It is at this point that fire prevention personnel all too frequently make a serious mistake that results in the invalidation of the evidence. Prior careful handling of evidence means nothing if this turnover is carelessly done.

B. Certain state officials, usually state fire marshals, are required by law to keep records of all arson fires and of other fires where criminality is involved. These records are not available to the general public unless specifically released by the appropriate authority. This should be borne in mind in report writing. It may be desirable to have two reports, one general in nature and open to the public, and the other a privileged report. Much can be said in a privileged report that cannot be said in a public report.

C. In most states, fire marshals or their equivalent have the authority to require assistance or information from other fire organizations, including fire prevention organizations. The extent of this authority differs from state to state. Providing such assistance or information is no problem if personnel assigned to suspected arson cases understand the situation.

D. All states have laws governing the right to manufacture, possess, distribute, or use explosives and certain other extremely hazardous materials. There are also federal laws on the subject. Lawful handlers of these materials have federal, state, or local certificates. If a person lacks such a certificate and possesses explosives or has in one area materials from which illegal fire bombs or explosives can be made, the situation constitutes prima facie evidence of intent to manufacture unlawful devices or to commit unlawful acts. Inspectors should be alert to such possibilities.

II. Making arson difficult. Primarily, of course, fire prevention authority is limited to making inspections and recommending remedial action. It is generally possible in very serious cases to call in legal authorities to compel correction. Within normal limits, however, proper inspections and recommendations help eliminate in advance conditions that make it easy for would-be arsonists. Such preventive measures include the following:

A. Through inspection services, eliminate unnecessary accumulations of hazardous materials.

B. Secure the installation of fire detection and protection systems where needed, and make certain they are properly maintained.

C. Educate management, employees, and others in fire safety, and secure their support.

III. Investigative procedures. When arson has been committed or is suspected, fire prevention personnel can do a great deal to prevent recurrences by proper conduct during investigations. Some of the methods used are as follows:

A. Collect and preserve evidence so that it will be acceptable in court.

B. Provide technical assistance to the prosecuting authorities.

C. Properly interrogate witnesses. It is necessary to remember that, from a fire prevention standpoint, it is the fire, not the real or suspected crime, that is being investigated. With that in mind, the inspector can ask any nonaccusatory questions that may shed light on the fire. Since the inspector is not a law officer, there is no obligation to warn the witness of the right to remain silent or the right to counsel. Once an accusatory question is asked, however, the witness must be given the Miranda Warning.

1. The Miranda Warning reads as follows:

a. You have the right to remain silent.

b. Anything you say can and will be used against you in a court of law.

c. You have the right to talk to a lawyer and have him present with you while you are being questioned.

d. If you cannot afford to hire a lawyer, one will be appointed to represent you before any questioning, if you wish one.

e. If you decide to answer questions now without your lawyer being present, you have the right to change your mind at any time, and request that your lawyer be present before you answer any further questions.

2. A waiver of the rights provided by the Miranda Warning must also be secured. After the warning and in order to secure a waiver, the following questions should be asked and an affirmative reply secured to each question.

   a. Do you understand each of these rights I have explained to you?

   b. Having these rights in mind, do you wish to talk to us now?

3. Other precauations. Fire prevention personnel who do their questioning properly should rarely need to use the Miranda Warning. In making interrogations regarding arson fires, however, there are several precautions that are necessary.

   a. If an interrogation is to be recorded, the fire prevention officer should so state at the beginning. This statement must include the date, the fire involved, and the name of the witness. Never let the recorder run on idle; gaps in the tape are questionable.

   b. In written records, identify the situation at the beginning and get the witness's signature if possible.

   c. If stenographic notes are taken, have the notes read back to the witness. Unless the witness is signing the report, this should be done in the presence of a third party.

IV. Autopsies. Many arson fires involve deaths. In general, autopsies are required by law for deaths in the fire area. Two points are of interest to fire safety personnel.

   A. Burns may cover stab wounds, broken skulls, or other lethal injuries. Make sure possible weapons are not carried out with the fire debris and that possible fingerprints are not lost.

   B. Deaths in an upper story may not be due to heart attacks, as so often reported. The coroner should be asked to check for toxic gas. If toxic gases are the cause of death, the person charged with arson may also be charged with manslaughter or murder.

V. Summation. In this preliminary introduction to arson law, the purpose has been to indicate in a general way how fire prevention personnel can, within the law, provide very useful assistance in curbing arson fires. Fire prevention organizations form part of a team that includes fire marshals, fire departments, police departments, insurance companies, and others. It is an unfortunate fact of life that, in far too many areas, these organizations have very little coordination with one another on matters of mutual concern. Fragmented efforts cost far more money and are far less effective than coordinated efforts. It is also a fact that arson losses drop immediately if each agency knows its responsibilities under the arson laws, has a working knowledge of the areas of concern of other agencies, and works with its partner agencies in a coordinated attack on the problem.

Arson Investigations

I. Goals. The proper way to begin any study of arson investigation techniques is, in the opinion of the author, to define what a good fire investigation is. A good investigation has the following characteristics.

A. It is a detailed attempt to uncover the facts and determine the truth. An investigation must be done in a logical order if logical conclusions are to be reached. An investigator must observe facts and ask questions so that he can piece together a chain of events, determine the cause, and attain any other pertinent objectives.

B. Every investigation should be conducted on the assumption that it might wind up in civil or criminal court. Most won't, of course, but that fact doesn't change the need to do each investigation correctly.

C. Whenever a fire is investigated, whether arson is suspected or not, the inspector should be interested in finding out some very definite facts.
 1. The point of fire origin.
 2. The source of ignition.
 3. Why a small fire became large.
 4. The reasons for casualties, if any.
 5. The extent of the loss, including business interruption. (Very important in the insurance field.)
 6. Why the fire deviated from normal, if that is the case. This factor is often essential in an arson case.

While it is true that a few fires start out large, the overwhelming majority start out small, even though a major fire may be in progress within a short time. An arson investigator should not lose sight of the fact that his primary objective is to determine the point of origin and/or the source of ignition. These facts are of critical importance in the successful prosecution of an arson case.

II. Fire behavior and fire patterns. The question is often asked, How can an investigator work back from a major fire to a point of origin, if there was no witness? To do this, an investigator needs a basic understanding of fire behavior and a working knowledge of typical fire patterns.

A. Structural fire behavior.
  1. Hot gases, including flames, are lighter than the surrounding air and rise. There must be a physical obstruction or a wind current of considerable velocity or its equivalent for fire to burn in any other way. A wind tunnel is one place where there are artificially induced winds of sufficient velocity to alter fire patterns. Lacking such abnormal conditions, a fire always burns upwards.
  2. Combustible materials in the path of rising flame ignite and increase the fire volume and the rate of upward movement.
  3. A fire can develop only when fuel is above the initial flame so that fire volume increases. Unless such fuel is present, the fire normally burns itself out. Exceptions to this rule of thumb are few and usually occur where special hazards are present. For example, careless handling might leave a downward trail of gun powder. This would flare in the event of fire and ignite combustibles below. It is probably safe to say that 98 percent of industry does not have this problem.
  4. Variation of the upward direction occurs when there are strong lateral air currents, downdrafts or other special conditions. For a flame to shift laterally, however, there still must be fuel ahead of the flame. Typical examples are as follows:
    a. Natural deflection. Some types of paint-drying operations that depend on natural ventilation may at times, because of local conditions, have strong downdrafts. This possibility can be pinpointed in advance by noticing the presence or absence of paint, varnish, or lacquer deposits on oven doors and where the deposits are.
    b. Induced deflection. At times, ventilating or air-conditioning systems are operated to create a downdraft at an escalator opening to aid rescue operations.
    c. Physical barriers. Lateral spread of fire increases rapidly when the upward spread is blocked by a physical barrier. If combustibles are present, the upward trend is resumed as soon as the fire finds a vertical opening.
  5. Flammable surface treatments. Downward spread occurs when highly flammable coatings or surface treatments are present. These materials burn off, and the fire behavior could be misinterpreted unless certain facts are recognized. For example:
    a. Assume a mock-up of such materials is ignited from the bottom. A standard burn pattern would result on the back-up wood, tile, or other materials.

b. Assume the same mock-up is ignited from the top. A markedly different burn pattern would be formed on the structural materials.
B. Fire patterns.
  1. Basic patterns. From the behavior characteristics cited above, certain conclusions can be drawn.
    a. Every fire forms a pattern determined by the environment, the type and availability of combustibles, and the presence or absence of installed fire prevention.
    b. A fire that follows basic patterns of behavior develops in the form of a more or less irregularly shaped, inverted cone. The apex at the bottom is the point of origin. In multistory buildings, each upper story will have one or more inverted cone formations, depending upon how many breakthroughs were made before the fires merged.
    c. When the fire pattern in the story of origin shows two or more inverted cones, there is every reason to operate on the assumption that the fire was incendiary in origin unless there is clear evidence to the contrary.
    d. The point of origin does not have to be a physical point, although the origins of most Class A and Class C fires come close to that. The point of origin of Class B fires is flattened into a small area, the size of which depends on the amount of free flammable liquid.
  2. Collateral information. Even though overall fire patterns may be modified by physical obstructions, fire load variations, outside weather, ventilating systems, and the like, the job of tracing a fire back to the point of origin in the event of a total or near total burnout is not quite as complicated as it might look at first. The following kinds of information are available and helpful.
    a. Fire department observations from time of response to extinguishment.
    b. Building construction details recorded in plans, inspection reports, and the like.
    c. Building utility data from the same sources.
    d. The fire load picture, assuming good fire inspection reports.
    e. The maintenance picture from reports.
    f. Last but not least, information from personnel who were present prior to the fire and had some knowledge of internal operations or witnessed some phase of the fire. Don't overlook fire photographs. They can be very revealing to the trained eye.

III. The structural investigation. Space does not permit a detailed review of arson investigating procedures, but the following brief outline of what is probably the most difficult type of problem for a new investigator can provide some useful general guidelines.

In this situation, a burning roof or floor collapses and deposits so much flaming debris that the original point of origin is obliterated. In addition, secondary fires start below the story of origin and perhaps collapse part of that story into the secondary fires. A story of origin does not usually collapse. Normally, even in a burnout, the lowest level to which the falling material goes is the floor of fire origin. In 90 percent of the cases, nothing below the level of origin is destroyed or damaged to the point where observations cannot be made. Falling material can, of course, cover the point of origin, but careful removal of debris will uncover this point. In perhaps 10 percent of the cases, floor openings permit burning debris to fall to a lower story, which may in turn burn out. Even in such a case, an educated guess is possible, granted knowledge of prefire conditions and good fire department observation.

In any burn pattern, low burn points should be checked out as possible point(s) of origin. In analyzing low burns, a rookie inspector may well make an error in interpreting what is seen, particularly if flammable liquids are assumed to have been present. Some of the potentials for error are discussed below.

A. If the fire originated on a tight floor, it is wrong to start with an assumption that flammable liquids were involved. Such a floor surface cannot be heated above the boiling point of the liquid, and this is below the ignition temperature of any floor likely to be used. As the last of the flammable liquid burns, some slight scorching may result if the liquid is gasoline or the like; there may be some discoloration if floor wax is present; or, with such liquids as alcohol, there may be no mark at all. For such fires, more information is needed before an origin in flammable liquids can be assumed.

B. While it is true that a flammable liquids fire tends to assume a rounded shape that is distinctly different from the ragged edges of most Class A fires, there are Class A materials, including many plastics, that melt before they burn and therefore produce rounded-edge burn patterns. Round, solid Class A objects may also leave rounded burn patterns. However, there are ways to distinguish among these materials of origin. Plastics, with rare exceptions, leave a noticeable residue, and Class A solids usually produce enough heat to cause fairly deep charring of nearby surfaces.

C. Although there may be no observable floor burn patterns, a wall or some other vertical or inclined surface may be observed burned to the floor level. This situation indicates the probable presence of flammable liquids, since even intense trash fires seldom burn a vertical member right down to the floor level.

D. Burning flammable liquids can follow floor cracks or floor-to-wall joints to a lower level and generate fires below. Most Class A fires cannot do this, because the materials involved can only drop through larger openings. The size of openings is of great significance.

By following procedures such as those outlined above, it is possible in a high percentage of cases to determine the point of origin, the source of ignition, and why a small fire became large. From these beginnings, it is also possible to determine whether or not there was any deviation from normal patterns and, if so, why.

IV. Summation.

A. Unless there is positive information on the fire cause, all potential sources of ignition should be checked out. This is important at all times, but it is vitally important when arson is suspected. In possible arson cases, it is essential that the inspector seek expert confirmation of findings outside of his or her training and education. For example, the inspector may know the cause was not electrical but lack an electrical background. This finding should be substantiated by a qualified electrician.

B. When a number of possible causes have been checked and ruled out, whatever cause or causes are left—improbable as they may seem —are the probable causes of fire, and the investigation should center on them.

C. If there is a potential arson case and all other causes have been ruled out by qualified personnel, the suspected arsonist is going to have a tough job to explain being in that area.

D. In checking out a possible ignition source, be sure that the check is complete. More than once, an inspector has made a serious blunder due to an incomplete check. For example, an inspector might be quite embarrassed if he listed the cause of fire as a short circuit in electrical wiring, and later it was proven that power for the building had been cut off a month before the fire. In arson cases, there is no room for guesswork.

Preparing Evidence

One of the most difficult crimes to prove and prosecute (not to detect) is arson. With arson—unlike such criminal acts as robbery, murder, and assault, where there is no doubt that crimes have been committed—it is first necessary to prove that a crime has been committed. Once that has been established, it becomes possible to determine Who? and Why?

Even if the fact of arson has been established, the investigator must assemble the evidence in such a manner that it can be presented in court and will convince a jury. The jury must be convinced, not only that arson has been committed, but also that the prosecuting attorney has valid evidence to support the charges against the person or persons on trial.

It is unfortunately true that far too few fire personnel have ever had any realistic training in the arson problem. As a result, many cases of arson go undetected and, in many others, evidence is so mishandled as to be inadmissible at a trial. In addition, many witnesses have had doubt thrown on their testimony because of lack of knowledge of how to conduct themselves.

I. Learn general patterns. Preparation for an arson trial should start long before the crime has been committed. In fact, it should start immediately after entry into the fire safety field, regardless of the branch entered. Study the fire loss picture as a whole, as well as individual losses. This kind of study shows the effect of fire on the business and social life of a community and how a plant can be shut down as the result of a fire in a supplying plant hundreds of miles away. It is not unknown for an arsonist to strike, not directly at the source of his grievance, but indirectly at some other person or operation that is important to the well-being of the community. Study of successful arson prosecutions will provide vital information on how to go about an investigation.

II. Learn how to find clues. Secondly, it is necessary to learn what motivates arsonists and how to find clues to their actions and motivations. Two basic guidelines are as follows:

    A. Look for something that doesn't belong. For example, clear evidence of flammable liquids or quick burning plastics in areas where they don't belong, either before or after a fire, should lead to further investigation.

    B. Look for something that belongs and isn't there. For example, a fire of any size in an office building might be expected to damage or destroy files and records, unless there was salvage during the fire. Removal of important equipment or records immediately prior to a fire could well raise some questions as to why this was done.

III. Learn arson law. The third step is to gain an understanding of arson law as applicable to the fire organization involved. Among other things, it is necessary to know the following:

A. What investigative responsibilities and authorities the fire organization has.

B. The relationship of one's own organization with other concerned agencies.

C. How to interrogate witnesses.

D. When an unexpected confession can be used.

E. How to take and preserve evidence.

IV. Obtain legally useful information. Normally, fire inspection and other fire safety personnel receive training in fire behavior, basic fire patterns, and potential fire causes. Since no one should be assigned to a suspected arson case without this background, for the purposes of this section it is assumed that the necessary background is available. Starting from this point, a number of types of information are critical to an arson case. Following are some basic guidelines to obtaining such information.

A. While it is always important to determine the fire cause, it is absolutely essential to do so in an arson case. Unless there is factual evidence that supports a conclusion of arson and rules out the possibility of other causes, the case is defeated before it starts.

B. Persons testifying on fire causes must be qualified in the field in question. (If an investigator rules out electrical wiring as a fire cause and is not a qualified electrician, the ruling should be corroborated by a person who is qualified in that field.)

C. The proper place to start investigating potential sources of a major fire is outside the building. Outside sources include power wiring and gas lines. Inside the building, start with major potential sources and proceed to the less important. The only exception to this procedure is when competent witnesses are available.

D. Fire department personnel should be trained in procedures to be followed to minimize the possibility of destruction of evidence when arson is suspected.

E. If there are fire deaths at a suspected arson fire, the autopsies should be conducted by an expert in forensic medicine. Coroners' autopsies that are standard for deaths in nonarson fires are usually not satisfactory for arson fire deaths, because the coroner lacks training in forensic medicine.

F. Unlike nonarson evidence, which may be discarded as soon as its bearing on the fire had been determined, arson evidence must be preserved for court trials. That is, it must be kept in protective custody as required by state laws so that the court can determine to its own satisfaction that there has been no opportunity for tampering. For example, records must be in bound books from which no pages have been removed.

G. Fire areas should be secured as necessary to prevent access that might destroy such evidence as footprints and tire tracks. An example of carelessness would be to permit an employee replacing the sprinkler heads to open a central valve by the wheel and smudge previous fingerprints. The value should have been opened by the spokes, pending a check on fingerprints on the wheel.

H. Amateur arsonists are not many and usually bungle so badly that ordinary police work catches them. The professional arsonist is another breed. Such persons understand fire behavior as well as the investigator. Once investigators recognize this, they start asking the right questions and checking the right things.

I. Erasures on written records are not permitted. If an error is made, line it out, insert the correct data, and have the change dated and initialed.

V. Become an effective witness. Usually, the attorney briefs the investigator on court procedures and how to conduct himself. The following guidelines may help an investigator be an effective witness.

A. Always present a neat and unruffled appearance.

B. Answer questions with quiet confidence in the correctness of your answers and in your professional competence.

C. Confine answers to fact and, as necessary, request court permission to refer to notes to refresh your memory.

D. Refuse to be browbeaten by an aggressive defense attorney.

E. Insist on the right to respond to questions in full.

F. Never be stampeded into volunteering opinions on subjects outside your area of competence. The defense attorney should if necessary be referred to an expert in the field in question.

VI. Summation. While it is quite true that the majority of fire prevention personnel will never get involved in an actual arson investigation, they should have a general knowledge of what is involved. When personnel have such knowledge, the inspector, the plan reviewer, and others through careful observation can do much to eliminate conditions that set the stage for arson. In addition, well-written reports of observations made during a prefire inspection often prove useful to the professional arson investigator. Concerted efforts in these directions will do much to curb arson losses and so improve the national fire loss records.

chapter 12
Plastics

Uses

Plastics have been around for a long time, but they have only come into widespread use in the years since World War II. In part the development of new plastics came about because of the wartime need to replace the supplies of natural products that were no longer available. It is probable, however, that the present huge plastics industry came into being primarily because of the utility and economy of the materials. Some major areas of use are listed below.

I. Clothing. The requirements for clothing the rapidly increasing population have far outstripped the capabilities of natural fibers to meet the demand.

II. Piping. Many highly desirable products and services depend on the ready availability of inexpensive, lightweight, easily assembled piping. Without such piping, the products would not be made, and the services would be curtailed or stopped.

III. Insulation. Many plastics have excellent insulating qualities plus light weight, easy workability, and low cost. These materials find an important place in building insulation, for tool handles where electrical insulation is needed, and for many similar purposes.

IV. Structural components. Low cost, light weight, and ease of assembly are among the characteristics that have made plastics widely used for windowpanes, wall panels, building siding, veneers, ceiling tiles, and other structural components.

V. Furniture. The characteristics of some plastics make them useful for seat padding, seat covering, table tops, and the like.

Characteristics

Unfortunately, most plastics are combustible, and some are seriously toxic under fire conditions. In addition, the manufacture of finished plastics almost without exception involves the use of large amounts of highly flammable materials.

Even though most finished plastics have only low to moderate flammability ratings, there are potentials for major fires and/or explosions when these materials are not properly handled. For these reasons, it is essential that fire prevention personnel have some understanding of the hazard problems that can be present in the manufacture, machining, and use of plastics.

Nitrocellulose

The logical place to start any discussion of plastics is with nitrocellulose, the oldest and, from the fire standpoint, the worst of the plastics. Contrary to popular belief, the advent of safety film did not mean the end of nitrocellulose in the public area. It is still used in considerable volume for coating fabrics and leather, in some fast-drying lacquers, in inks, for paper coating, for tool handles, and for many other processes and materials.

I. Manufacturing process. Cellulose (usually cotton or wood pulp) is dipped in concentrated solutions of nitric and sulfuric acids. To vary the end properties, the strengths of the acids, the length of dipping time, and the proportion of cellulose to acid are modified. Following dipping, the excess acids are removed, and the resultant product is washed, dried, and pulped. The end product of these processes is nitrocellulose.

II. Product characteristics.

  A. Nitration. During the processes cited above, various percentages of nitrogen and free oxygen are added to the cellulose molecules. The amounts of nitrogen and free oxygen joining each molecule depend on the process variations described above. It is important to note the following facts.

  1. The higher the percentage of nitrogen, the more dangerous is the material.

  2. Because free oxygen is built into the nitrocellulose molecules, attempts to extinguish fires by excluding air are useless.

Plastics

B. Burning rate. Bulk nitrocellulose storage with no installed fire protection burns at the rate of approximately 1,400 lb. per min. The substance generates fewer BTUs per lb. than most other combustibles, but its tremendously fast burning rate means that, anywhere nitrocellulose is stored in bulk, there is the potential for an extremely fast fire with the evolution of tremendous heat.

C. Stability. Nitrocellulose decomposes at relatively low temperatures and evolves large volumes of highly flammable and toxic gases.. Temperatures must be kept below 100° F. and preferably below 90° F. to prevent self-sustaining decomposition.

D. Plasticizers. Various materials called plasticizers (camphor is one) are added to nitrocellulose to assist in preventing decomposition and to produce other desired properties.

III. Transportation requirements. Cellulose nitrate may be shipped wet, in which case the containers should have a DOT yellow flammable solids label. The substance may also be shipped with a wetting agent of flammable liquid, in which case a DOT red flammable liquid label is in order. Shipments that are intrastate do not come under DOT supervision, but most states require similar labeling. Where local regulations are lacking, intrastate shippers should be encouraged to use standard warning labels.

IV. Storage requirements.

A. Quantity. Many codes specify the maximum amount of nitrocellulose that may be stored in a given area. Code requirements also cover sizes of piles, vault and cabinet requirements, and room or building construction.

B. Heat. Steam, hot water, and electricity are the preferred means of heating. Systems should be equipped with temperature-limiting devices so that temperatures will be kept at a level below that at which decomposition takes place. Similar care must be taken at hot processes in manufacturing operations.

C. Ignition sources. These must be eliminated. For example, fixed lamps should have vapor-tight globes.

D. Electrical equipment. All such equipment must comply with the *National Electric Code.*

E. Occupancy separation. Storage areas must be separated from other occupancies by appropriately rated fire walls and floors.

F. Fire protection. For appreciable nitrocellulose storage, sprinkler systems should be designed to deliver from 1 to 1.5. gpm per sq. ft. of floor area. Because sprinkler heads must be closely spaced, they should be separated by baffles. Baffles shield the soldered links of sprinkler heads from water from nearby operating heads. Water on the soldered link would of course delay a head's action.

G. Explosion venting. Provision should be made for venting the pressures that would be generated by the ignition of decomposition vapors.

H. Decomposition venting. Despite all precautions, some decomposition must be anticipated, and ventilation must be provided for the products of decomposition.

V. Manufacturing operations requirements.

A. Water should be used as a coolant wherever saws, grinders, drills, and the like are used.

B. Smoking and the carrying of matches must be prohibited.

C. Ignition sources must be eliminated to the extent possible. Where this cannot be done, the plant must install such safety equipment as heavy wire guards over ignition sources, baffles or shields between processes and ignition sources, and safety lubrication to eliminate mechanical sparks.

D. Scrap nitrates should be stored wet in metal containers and should be removed daily (more frequently if necessary) from the building for disposal in a safe area.

E. Automatic sprinkler protection should be provided as required by the provisions of NFPA Standard 13.

VI. Summation. The outline furnished above provides basic guidelines on the hazards of cellulose nitrate. Inspectors should be alert to the possibility of its being present in many areas. Where it is found, care should be taken to see that it is properly housed, used, and protected. Additionally, it is desirable to make certain that responding fire departments are aware of its presence. Cellulose nitrate fires are often quite spectacular and can present severe fire, explosion, and toxic hazards to both fire service personnel and employees. For this reason, the author believes that, in addition to standard protection systems, the provision of an evacuation alarm is desirable. Such alarms would be triggered by sprinkler operation or by devices sensing advanced decomposition.

Basic Classifications

There are two basic types of plastics: thermoplastic and thermosetting. Fire prevention personnel should be particularly aware of the difference between the two, as this difference can be very important in determining actual or potential losses in the event of fire.

Plastics

I. Thermoplastic materials. Thermoplastic compounds are softened by heat and hardened by cooling. No chemical change takes place as a result of these processes, and the materials can be broken up and reused repeatedly. Because of this capability, thermoplastic items that have been melted or distorted by heat from fire are, unless contaminated, salvageable for reuse in producing new products. Normally, thermoplastic materials contain no fillers or binders. They are principally used for products made by injection molding, extrusion, or casting.

II. Thermosetting materials. Thermosetting compounds normally include fillers and/or binders in addition to the raw plastics. All these materials are mixed together under heat and pressure. The mixture softens and can be used to fill a mold. While this process is going on, a chemical change takes place. Because of the chemical change, the mixture, when it has cooled and hardened, is a finished product. It cannot be broken up and reused. Therefore, when finished thermosetting plastics products have been damaged or distorted by heat from a fire, they are, for all practical purposes, a total loss.

III. Summation. With the two types of plastics and their innumerable combinations, industry can tailor-make a plastics material to fit almost any need, from building siding to foamed insulation, tool handles, piping, and toys. These capabilities should be used, but used with judgment. Wisely used, plastics are an efficient and economical means to a higher standard of living. Wrongly used, plastics can contribute to disastrous fires and to loss of life in minor fires.

Creating Plastics

I. Hazards.

A. Raw materials. In the creation of plastics, it is well not to lose sight of the fact that many of the basic raw materials are highly flammable. Many of them also have other interesting properties, such as a tendency to explode, toxicity, corrosiveness, ability to contribute to a runaway chemical reaction, and, in a few cases, radioactivity. Some of the more commonly used basic materials and a few of their important characteristics are shown in Table 9.

Some of the materials listed in the table were previously cited in Chapter 10.

B. Processing. In addition to hazardous materials that are used as basic raw materials, the processing, depending on the end product, may include such hazardous materials as those listed below.

1. Corrosive materials.

Table 9
Characteristics of Raw Materials for Plastics

	Flash Point	Explosive Range	Selected Hazard Properties
Acrylonitrile	32° F.*	3.0–17.0	1,3,4,5
Butadiene	−105° F.†	2.0–12.0	3,4,5
Ethylene	−185° F.†	2.7–34.0	3,4,5,7
Formaldehyde	Gas	7.0–73.0	3,4,6,7
Propylene	−162° F.†	2.0–11.0	3,4,5,6,7
Styrene	90° F.†	1.1–6.1	2,3,4,7

1 = Severely toxic. (All of listed materials do have some degree of toxicity.)
2 = Hazardous in contact with combustible materials.
3 = Hazardous when heated.
4 = Highly combustible.
5 = Hazardous with oxidizers.
6 = Explosive when mixed with air.
7 = May explode in fire.
* = Open cup flash point.
† = Closed cup flash point.

2. Fillers, many of which are combustible and some of which are in powder form. Some horrendous explosions have resulted when combustible filler powders were accidentally dispersed in the air.

3. Miscellaneous flammable liquids, gases, and solids other than the base materials.

4. Organic peroxides.

C. General. The listings previously cited name only some of the hazards that may be present in the creation of plastics. Literally tens of thousands of possible combinations of flammable liquids, flammable gases, natural and synthetic resins, combustible dusts, and the like can be used to create plastics. From the inspection and plan review standpoints, the only practical solution to the problem is to review each operation on an individual basis, taking into account the materials actually used or proposed for use in a specific process. When the problem is brought down to those terms and there is a general understanding of the characteristics of the materials used, the situation becomes manageable.

II. Controls. The creation of plastics is primarily a chemical process and involves the use of reaction vessels, autoclaves, stills, and allied equipment. During the processing, there may be and often is a possibility of a runaway chemical reaction, and such reactions have resulted in fires and explosions. To prevent such incidents, it is vitally important that controls be of the proper type and that they be given top quality maintenance. There should also be a program for equipment maintenance to forestall the possibility of leaks. Virtually all of the serious fires and explosions that have occurred at this stage of processing have occurred from one of three factors: (1) excessive temperature or pressure, (2) leaks, and (3) poor control of mixing.

Much of the hazard problem can be removed by the proper selection and training of employees. As required by their work, employees must know the following:

A. Correct procedures for controlling temperatures and pressures, including the use of relief valves when a reaction is proceeding too rapidly.

B. When and how to add ingredients to the mix.

C. How to control and use inert atmospheres.

D. Standard fire prevention measures, such as control of ignition sources, the operation of fire protection systems and equipment, housekeeping requirements (with specific reference to dust control), and the like.

III. Life safety. Normally, processes where a serious runaway chemical reaction is possible or where dust explosions could occur are housed in buildings designed to vent explosive pressures. (In some cases, the buildings have open walls.) Special construction does go a long way towards limiting property damage, but in itself it may do little to assure employee safety. In the author's opinion such buildings should also have the following equipment.

A. A visible and/or audible alarm system designed to operate before critical conditions are reached.

B. Safety evacuation chutes or their equivalent to assure prompt evacuation.

IV. Storage.

A. Raw materials should be stored in accordance with the requirements previously noted for flammable liquids and gases and for special hazard chemicals. (See chapters 9 and 10.)

B. Finished stock storage should be stored under the following conditions.

1. Storage areas should be separated by fire resistive barriers from both raw stock storage and manufacturing operations.

2. Height, size, and separation of piles should be determined by the published standards of the authority having jurisdiction. Where no code requirements have been published, plants should be encouraged to comply with the published recommendations of

the NFPA, the Factory Mutual System, or another nationally recognized authority. All the developed standards base the limits on the degree of combustibility and whether or not the materials have been foamed.

3. Stored finished stock should be separated from walls and unprotected building columns by clear spaces that, if possible, are 3 ft. or more across.

4. Areas for finished storage should have automatic means for venting smoke and heat.

V. Fire protection.

A. Raw stocks should be provided with the same type of fire protection that would be provided for similar materials in other types of manufacturing operations.

B. Preferably, finished stocks in quantity should be protected by wet pipe sprinkler systems. Where this is not feasible, dry pipe or preaction systems are acceptable. In any case, the systems should be designed to deliver a somewhat higher volume of water than would be provided for ordinary combustibles.

Finished Product Hazards

Plastics are finding an ever-expanding place in building construction, decorative finishes, household and public assembly furnishings, and articles for personal use. In finished form, plastics have flammability, toxicity, and reactivity ratings that are substantially lower than those for raw stocks. Despite these reductions in hazard ratings, a number of serious fires and explosions have involved finished plastics. Some of these incidents have included plastics that have been given fire retardant treatment.

In the short space of one chapter it is not, of course, possible to detail all the potential hazards, but the following comments and examples provide a basic framework on which to build.

I. Foamed or expanded plastics. Foamed plastics are produced by mixing a gas (usually air or carbon dioxide) with the plastics materials while they are still in a liquid state. The end product is a cellular structure of extremely light weight. The product may be rigid or quite flexible, depending on the materials used and the overall processing. Foamed or expanded plastics have the following uses and characteristics.

A. Used for building insulation, packing materials, padding, seat cushions, mattresses, and many other products.

B. Flammable, and once ignited tend to burn rapidly with the evolution of dense smoke. In general, even those treated with fire retardant materials burn, although ignition takes longer and the burning rate is somewhat slower.

C. Beginning to be combined with nonplastics to form building-insulating panels that have been approved by major insurance carriers.

II. Dusts. Although many plastics are difficult to ignite when in the form of large solids, nearly all burn with explosive rapidity when in the form of dusts dispersed in the air. (Many plastics dusts are capable of generating explosive pressures of from 65 to 100 psi.) It is therefore important to prevent the accumulation of dusts by setting up operations to minimize the production of dust and to provide for its regular removal.

III. Static electricity. The same properties that make plastics good electrical insulators also enable them to rapidly build up static charges. This characteristic makes it necessary to provide appropriate means for eliminating hazardous static buildups where combustible dusts, flammable liquids and gases, or other highly combustible materials may be present. Measures may include equipment grounding and/or appropriate static eliminators. It should be noted that, unlike most other combustibles, plastics build up static regardless of the humidity.

IV. Melting. Most plastics cannot stay in service at temperatures above 300° F. Some have a limit of 250° F. or less. When a plastics material reaches its working limit, it tends to distort and, in many cases, starts to drip. These temperature requirements put a definite limit on how plastics can be used. Those that drip have limited use overhead.

V. Toxicity. All plastics under fire conditions emit toxic gases. Gases from certain plastics may be lethally toxic. Gases from plastics at the other end of the scale have relatively low toxicity ratings. However, none should be ignored.

VI. Corrosion. Most vinyl plastics under fire conditions generate hydrogen chloride. This gas usually spreads well beyond the fire and is capable of damaging human tissue as well as metal objects even at remote locations.

VII. Textiles. The fire behavior of flammable plastics textiles is similar to that of natural-fiber textiles in that the burning rate is determined by the type of weave and the length of nap. Burning clothing made of plastics can create problems in addition to normal burns because some plastics tend to adhere to the skin. Most plastics textiles have been treated so that they may safely be worn. A small percentage of plastics fabrics is highly flammable, but such fabrics are seldom used for wearing apparel.

VIII. Fire retardance. Many plastics are treated to slow down their burning rates. Unfortunately, all such fire retardant treatments do not have the same degree of permanence. The important question here is, If the protected material has once been exposed to a fire or heat from a fire, how much fire retarding capability is left for a possible second fire? The answer to this could be of critical importance in selecting plastics for use with an important computer installation. Fire retardance capability is also of major importance to the processor who manufactures various articles from the finished plastic materials, to the user, and to the fire prevention inspector.

Fire Tests

Plastics, particularly those used in the building industry, have been subjected to tunnel tests in the same manner as other building materials. These tests provide flame spread, fuel contribution, and smoke emission ratings. For most combustibles, such information about a material's fire behavior holds true even when the material is installed in different manner than in the test. The same is not true for many plastics, particularly foamed plastics. For example, foamed plastic panels in a vertical position burned far more rapidly than the same material did in a tunnel test. This fact was proven in the Factory Mutual corner test. This differential is now recognized, and joint efforts are being made to develop more realistic tests. This does not mean that current ratings should not be used. It does mean that they should be used with caution.

Some Instructive Fires

I. Airport terminal building. About two years ago, a fire resistive building to be used as a passenger lounge and for ticket sales was under construction at a New York airport. The building was nearly finished, and workers had already installed various plastics materials for decorative effect as well as seats with synthetic coverings over foam rubber cushions. A fire started and, despite the efforts of the nearby excellent airport fire department, spread through the lounge area. Damage was in excess of $2 million.

II. Apartment house. Fire, probably from a carelessly discarded cigarette, moved slowly down a hallway carpet in an apartment house. Fumes from the burning plastic fibers entered the open door of an apartment. The occupant delayed too long in evacuating and died from inhaling the toxic fumes. The fire was never large and did only minor damage to the hallway.

III. Hospitals. During the late 1960s, a certain type of carpeting was proposed for installation in a number of areas of a large hospital then under construction. Small-scale fire tests of carpet samples showed that these particular plastic fiber carpets burned vigorously and emitted large volumes of dense smoke. The builders immediately substituted carpeting having a flame spread rating below 25 and a low smoke emission rating.

IV. Correctional institution. In June, 1975, a fire occurred in several stacks of urethane mattresses, which, incidentally, had covers treated for fire retardance. Those mattresses were stored in a room adjacent to prisoner cells. The rapid buildup of heat and toxic gases resulted in the deaths of ten prisoners and one correctional officer before rescuers could reach them. While there were other factors involved, it is probable that the major cause of this tragedy was the lack of understanding of how fast a urethane fire can build up. Had there been such an understanding, it is reasonable to assume that the mattresses would have been stored elsewhere under sprinkler protection and that prison design would have provided for prompt evacuation of prisoners exposed to a potential urethane fire. Do not be misled by the failure of the fire retardant treatment. On individual mattresses it would have been effective, but no retardant treatment for a cover is designed to resist a mass fire.

Summation

As has been pointed out, plastics are made or processed in many communities and found to some extent in nearly every area where people live, work, or play. It has also been noted that there are thousands of varieties of plastics, with new ones coming on the market every day. The very number of varieties guarantees that the products will have widely differing degrees of flammability, toxicity, reactivity, corrosiveness, and other characteristics.

Fire prevention personnel are faced with the need to develop realistic means to deal with the problem. Obviously it is impossible to learn and remember all the details of tens of thousands of products. It should also be obvious that the wrong place to learn is at the fireground. What must be done is to reduce the problem to a size that can be dealt with. Some of the means to do this are discussed below.

I. Identification. Fire prevention personnel can learn the general classes to which the various trade named plastics belong. These classes, including polyvinyls, styrenes, polyesters, and urethanes, all have definite parameters of characteristics and types of usage.

II. Safeguards. The materials belonging to each basic class of plastics require approximately the same type of fire protection and safety controls, given the same type of processing, storage, or use.

III. New plastics. The possible toxicity of each new plastic should be viewed with suspicion until proven innocent.

IV. Fire retardant and self-extinguishing classes. Do not rely solely on tunnel tests. If possible, check on availability of large-scale tests of materials in position as proposed for use.

V. Handling. Much is still unknown about plastics behavior, but one point must be emphasized and reemphasized. Plastics have properties that must be respected. Mishandling is a major cause of many large plastics fires.

In summary, preplanning to prevent serious plastics fires requires developing a workable system of identification, learning the types of products composing each major class, gaining an understanding of appropriate fire protection and safety controls for each class, and learning proper handling procedures for plastics in quantity in the areas being serviced.

chapter 13
General Occupancy Problems

Occupancies range from steel manufacturing to schools, from oil refining to grocery stores, from textile plants to furniture manufacturing, but all have one feature in common. With rare exceptions, disastrous fires and/or explosions are caused by the neglect of fundamentals and not by unusual, complex situations. While it is true that special hazards have at times played a major role in serious fires, it is also true that the primary reason these hazards became involved in a fire was neglect of fundamentals.

The above conclusion can be easily substantiated by careful reading of published fire reports and fire safety articles of recognized authorities such as the NFPA and the Factory Mutual System. Such data can also, as desired, be supplemented by similar information appearing in the press and certain magazines such as *Reader's Digest*.

Fundamental occupancy problems are few in number and not too difficult to understand readily. These basic elements should be securely grasped and applied to all kinds of occupancies.

Housekeeping

Of all ways to prevent fires, housekeeping is the most important. There can be no fire, nor can there be any major spread of fire, unless combustibles are present. There has to be something to burn for fire to start or, once started, to spread. Some of the major factors involved in a satisfactory program for general order and neatness are outlined below.

I. General housekeeping. Too often, the initial fuel source consists of accumulations of lint or other combustibles on machines and motors; combustible debris in pipe trenches; excessive combustibles under work benches; trash in open and/or combustible containers; or the like. In laundries, for example, debris tends to accumulate in the gears and motors of mangles, in the exhaust setup behind tumblers, in lint traps, and in similar places. An inspector should always keep plant operating hours in mind when checking for such debris. It does no good to check immediately after a weekend maintenance shutdown.

II. Waste collection and disposal.

A. In general, waste containers should be metal with self-closing covers. In a few kinds of occupancies, covers should be tight closing.

B. A few kinds of scrap materials should be stored under water. This should be a matter of record.

C. Waste materials should be removed from operating areas at least daily. Removal should be more frequent if necessary to prevent excessive accumulations of combustibles.

D. Waste receptacles should be in accessible places, but under no condition should these containers be placed in an exitway or near a vertical opening. (The location of trash containers in a stairway was a major factor in a tragic school fire that took the lives of nearly 100 children.)

E. Trash collection areas should be in suitable rooms that are cut off from other areas or in other safe locations. If Dempster Dumpsters or their equivalent are used, locate the containers so that, in the event of fire, building contents will not be exposed.

F. Wastes should be disposed of by such means as burning in a properly located and designed incinerator or removal to a commercial disposal facility.

G. In many plants, dusts and other light items such as wood shavings are removed by a pneumatic collecting system that discharges into a collector or storage bin. Preferably, collectors and storage bins should be outside, but some are indoors. Some points to remember are the following:

1. Removal of combustible collections must be on a regular basis.
2. Larger types of collectors or, for that matter, small important units should have installed fire protection.
3. Where magnesium, aluminum, titanium, or similar metal dusts are collected, collector must be of the wet type.
4. Where rags, waste, nails, or other foreign material might enter the system, suitable screens should be maintained over duct inlets.

(For example, floor sweep inlets in carpenter box shops should be screened.)

H. Virtually all organic materials, including charcoal, linseed oil, and animal oils, tend to heat spontaneously. This tendency is increased by high temperatures. It is therefore important that such materials not be allowed to accumulate on such hot surfaces as steam pipes.

III. Smoking. In spite of numerous "No Smoking" campaigns, year after year careless smoking is responsible for a large number of fires and substantial losses in lives and property. In the writer's opinion, the principal reason this is so is a generally wrong approach to the problem. For example, assume a hypothetical plant manufacturing screws, bolts, and nuts in noncombustible buildings. Except for shipping and receiving areas and certain stock rooms, most of the plant contains virtually no combustibles other than wood skids and tote boxes. Next to the manufacturing building stands an office building of ordinary construction. It has considerable wood wall paneling and wood furniture. Office supplies in quantity are in the open. The plant has a "No Smoking" policy for the factory but allows smoking in the office. How well does such a policy work?

John Doe, who works in the factory, visits the office and sees clerks, secretaries, and other personnel surrounded by piles of paper; many are smoking. His reaction is predictable: If management can smoke surrounded by paper, there is no reason for me not to smoke around noncombustible machinery in a noncombustible building. Therefore, he smokes; but because there is a "No Smoking" rule, the cigarette is tossed under a machine, into a tote box, or elsewhere whenever management is spotted. From such situations spring many fires.

Arbitrary "No Smoking" rules are both unrealistic and unenforceable. It is far better to control smoking than to try to completely prohibit it for the simple reason that control works. Some of the basic provisions for a realistic control program are listed below.

A. Absolutely prohibit smoking in areas where flammable liquids and gases, combustible dusts or lint, or similar materials are present. No one would seriously question "No Smoking" rules in such areas, and the regulation is easily enforced.

B. Allow smoking in noncombustible processing areas, unless smoking might contaminate the product. If ash trays are available, butts will not be hidden under equipment or in similar places, and fires will no longer result from such disposal practices.

C. Where smoking must be prohibited, provide designated smoking areas with reasonable access thereto. Such provisions usually assure good employee cooperation.

In short, smoking regulations should be set up on the basis of what is realistically needed to prevent fire and for no other reason except as required by health authorities.

IV. Maintenance. While not specifically housekeeping, preventive maintenance is closely enough related to housekeeping to warrant inclusion under this heading. In previous chapters, mention has been made of correct maintenance practices for various types of equipment, and this information will not be repeated here. There are, however, some other preventive maintenance practices that have general application. Among these are the following:

  A. Schedule maintenance activities so that there is a minimum of interference with production.

  B. Train maintenance personnel in advance to observe plant safety precautions.

V. Safe operations. Applicable to all operations are the following provisions to enhance operating safety.

  A. Whenever an operation is to be expanded, it is best to split the original trained crew between the two operations and flesh out both units with new employees.

  B. The training program should include provision for full training of new employees and periodic refresher training for experienced employees.

Because of the increasingly automated state of manufacturing, the potential for damage from human error grows larger each year. For this reason, the above provisions will doubtless assume even greater importance in the years to come.

Key Operations

The function of a well-run fire prevention organization is to prevent loss. Eliminating fire causes to the extent possible is not enough; neither is tight control of property values exposed to a single fire potential. Despite all efforts in these directions, there will inevitably remain some potential fire causes that cannot be completely eliminated and some large value concentrations that are essential to efficient industrial operations. In the present state of the art, 100 percent success in eliminating fires is not possible.

Since there will inevitably be some fires, fire prevention programs must be geared to looking beyond direct property losses to potential serious losses far beyond the immediate scene. When this area is surveyed, it quickly becomes evident that, frequently, indirect losses far exceed direct property fire losses. Fire prevention personnel should therefore be trained to look for operations that are essential to the continuity of production, both at the potential fire scene and elsewhere.

Nearly every industrial plant, institution, commercial establishment, or other enterprise has one or more key operations the loss of which would cripple the entire undertaking. Fire prevention personnel should be alert to these possibilities and to ways of minimizing loss should a fire occur. Some examples of key operations are as follows:

I. Water supply. A community's water supply requires chlorination to be potable. Chlorination is done automatically in an unattended, unprotected, wood frame building. Lightning strikes the building, and it is destroyed. Pending the construction of a new facility, the community is without drinking water. Even after rebuilding is completed, the water distribution system requires flushing. A disease epidemic might even have occurred before the fire damage was discovered. Direct fire damage might be no more than $5,000. Cost of a new facility, trucking in water supplies, medical costs, and cleaning the distribution system might well run $75,000 or more. If plants in the community shut down for lack of potable water, costs would be higher still. Sprinkler protection for a small building costs perhaps $2,000. But too many inspectors may not recommend a sprinkler system because they see only direct property loss.

II. Computers. In today's increasingly automated world, many operations are run by computers. It is also more and more common to handle records by computer systems, thereby eliminating paper records. A computer fire may cause losses far beyond the value of the computer, great as that may be. For example:

A. A steel plant worth perhaps $500 million is operated by computers. Shutting down a blast furnace or certain other operations may take two or more days. Starting up also takes time. If the computer controlling such operations is knocked out by a small fire, is there computer capacity in reserve to take over the load? If not, how well is the plant equipped to operate under manual control while the computer is being repaired or replaced?

B. Computers place important data on tapes or magnetic discs. A relatively small fire can destroy tape or disc memories. If this happens, are there duplicate tapes or discs? Assuming there are duplicates of important records—and there should be—these duplicates should be stored in a separate fire resistive building, room, or vault.

III. Air conditioning. At one stage of the manufacturing process, the product must pass through a dust-free, controlled humidity atmosphere. A small fire destroys combustible filters and their housing at a loss of $3,500. The plant is shut down while the filter system is being replaced, and the cost of lost production totals $100,000. The plant shutdown could have been avoided had there beeen a better choice of filters or proper protection for combustible components.

IV. Heating plants. A 500-bed hospital in a cold climate is served by a single oil-fired boiler. A backlash knocks out the burner when the outside temperature is $-30°$ F. Estimate the cost of moving critically ill patients to other facilities—not to mention the likelihood of losing some of them.

V. Water-sensitive products.

A. Water-sensitive crystals are stored in a room below a hazardous operation. Floor between is not waterproof. A small fire in the hazard area is quickly extinguished by sprinklers. Water, however, leaks through the floor and destroys nearly half a million dollars' worth of crystals. In addition, the plant is shut down for weeks.

B. Tenant in top story of a four-story joisted building has a sprinkler protected painting operation. A tenant in the first story has television sets stored in corrugated paper cartons on the floor. First tenant has a fire that is extinguished with minor damage by three sprinkler heads and a few minutes' use of a hose line. Leakage to the first floor wets the bottom tier of cartons. These cartons fail, and the piles collapse. Approximately $100,000 worth of television sets are destroyed. Watertight floors or first floor skids would have prevented this loss.

VI. Critical components. Perhaps the classic example of how a fire in one location can create losses in other areas is the General Motors fire in Livonia, Michigan, on August 12, 1953. The direct property loss was about $50 million, but this was only the tip of the iceberg. The destroyed plant manufactured all the automatic transmissions for several major makes of automobiles. There is no way to estimate indirect losses which, in addition to assembly plant shutdowns, unquestionably seriously affected plants supplying parts to the assembly plants. Inevitably, unemployment funds were drawn on, and so the circle widened. Because General Motors had a highly qualified staff in all areas of concern, and exceptional resources, the corporation rapidly overcame the effects of this loss. It should be noted, however, that not too many organizations have those kinds of capabilities, and other companies might find it extremely difficult to recover from such an event.

General Occupancy Problems 257

These seven cases are only samples from fire loss history. More often than most people realize, fires large or small at key locations have or can have serious effects beyond their immediate area. Fire prevention personnel should be alert to such situations.

Plant Shutdowns

Periodically, virtually every industrial plant is shut down for such reasons as retooling, strikes, vacations, or inability to get supplies. Nonmanufacturing facilities are also shutdown from time to time. During these periods, there are few employees on the premises. It is therefore of utmost importance that fire safety precautions for these periods be carefully preplanned.

I. Typical precautions. Some typical precautions to be taken when a plant shuts down are outlined below.

    A. A close-of-business inspection should be made by qualified employees within the first hour after operations have ceased. Among other things, these employees should check to see that

        1. all equipment except that required for essential services and life tests has been shut down;

        2. all sprinkler control valves are open; and

        3. possible fire sources have been removed or adequately safeguarded against.

    B. The fire department should be notified. This notification should include the names and locations of individuals to be alerted in event of an emergency.

    C. Schedule advance tests of fire protection systems, boiler combustion safeguards, and other safety equipment.

    D. Provide such special instructions as may be necessary for any skeleton force on duty.

    E. Any work to be performed by outside contractors should be supervised to the extent necessary to assure compliance with plant safety rules.

    F. Provide special security when needed. For example, important records and plans in a closed and essentially unattended plant should be in appropriate vaults or fire resistive cabinets. Key files should be identified by a special marking known to the fire department, so that they would be removed first in the event of fire.

    G. Curtail deliveries of highly flammable materials that are not immediately needed. These deliveries should be scheduled for after plant reopening, when a full staff will be on hand.

II. General goals. No two plants are alike, and no two necessarily

need the same plant shutdown fire safety program. All, however, have the same basic requirements which are listed below.
  A. Eliminate to the extent possible potential fire sources, and provide safeguards where sources cannot be eliminated.
  B. Limit the amount of combustibles to the minimum possible.
  C. Maintain fire protection systems.
  D. Notify fire department.
  E. Provide inspection, training, and supervisory services as necessary.

Disaster Recovery

There is no such thing as a plant or an institution or any other human facility that is immune to disaster. No matter how well situated a plant or facility is or how well protected it may be, it can still be struck by earthquake, tidal wave, flood, windstorm, outside conflagration, radioactive material incident, or another natural or man-made catastrophic force. Offhand, most people would probably question the inclusion of some of these forces in the field of fire prevention. They do belong, however, because many major fires have occurred collaterally with each one of these forces. The only way to avoid such losses is to preplan for possible emergencies.

I. Floods. Thousands of industrial plants, warehouses, and other types of structures are located in floodplains or in areas subject to occasional flooding. Each year, a sizeable number of these structures are flooded, and each year some have fires while flooded. Whether these fires are major or minor depends to a large extent on the measures taken in advance. Some precautions to take when flood threatens are listed below.
  A. Move vehicles and tank cars containing flammable liquids or gases, chemicals that react with water, or other hazardous substances to higher ground.
  B. Keep automatic sprinkler systems in service. Do not forget that, in the event of fire, normal fire department response may not be available. More than one plant has been destroyed by fire because the sprinklers were shut off and flood waters prevented the fire department from reaching the plant.
  C. Shut off gas supplies at the main control valves. Flood waters often break distribution gas piping and release gas unless these valves have been closed.
  D. Cut off electric power from areas likely to be flooded.

General Occupancy Problems 259

E. Fill vertical tanks of flammable liquids with water to prevent their floating away.

F. Close discharge valves of tanks and other containers of flammables.

G. Extinguish ignition sources. In some instances, this may mean pulling boiler fires.

II. Earthquakes. Unlike floods, earthquakes do not announce themselves in advance. These catastrophies may release combustibles, break fire cutoffs, and do much other damage, and so it is almost inevitable that fires will occur in the aftermath of a serious earthquake. Many of these fires can be controlled with minor damage if the following precautions are observed.

A. Sprinkler systems in earthquake-prone areas should be designed with earthquakes in mind. Experience has proven that systems provided with flexible couplings and sway bracing where needed usually survive all but the most severe quakes.

B. Gravity and pressure water tanks in seismically active areas should be designed to resist lateral earthquake forces. The importance of this precaution may be seen in the fact that, in the event of a severe quake, the public water system is usually a casualty. If the tanks fail too, there is no water to fight fires.

These precautions, as well as others pertaining to gas piping and other systems, are required on much of the West Coast. It must be remembered, however, that other parts of the United States also have a record of seismic activity, and some plants in these areas should take similar precautions because of the nature of their operations. Incidentally, it was the fires following the earthquake and not the quake itself that destroyed San Francisco in 1906.

III. Winds and waves. In the author's files are pictures showing fire damage and other destruction resulting from the New England hurricane of 1938. In southern New England, hurricane winds were acompanied by unusually high tides.

During the peak of the storm, many fires were started by broken power lines or other hazards. Some of these fires were of major proportions. One in New London, Connecticut, for example, destroyed over a quarter of a square mile in a heavily built-up area. Many of the major fires occurred because sprinklers were shut off or not installed where needed or because fallen trees, flooding tidal waters, and other barriers prevented fire departments from mounting effective attacks.

New England learned a great deal from this disaster. Many code changes were made, and many plant programs were greatly improved. However, many coastal areas remain wide open to similar disasters.

IV. Outside conflagrations. Not too many years ago, there was a sprinklered woodworking plant situated in the middle of a somewhat wet meadow. In the yard were moderate stocks of lumber protected by a system of yard hydrants. It was suggested to the plant that arrangements be made to have the meadow grass for a reasonable distance around the plant cut and removed. For whatever reason, the grass was not cut. During the following months, there was a drought, and the meadow dried out. The dense, high, dry grass became ignited, and the ensuing fire rolled into the plant. Before the fire was extinguished, much of the yard storage was destroyed, and the plant itself was seriously damaged.

V. Summation. Not all plants are susceptible to serious fire damage in the event of a natural disaster, but many are. Fire prevention personnel should be alert to the possibility of a natural disaster that could result in serious fires. Where this is possible, every effort should be made to make sure that proper preplanning sets up effective safeguards against fire. It should be remembered that a major fire disaster on top of a natural disaster may make it impossible for a business organization to survive, whether or not the loss is covered by insurance. Fire preplanning ensures the survival of many companies that would otherwise go under when hit by a natural disaster.

Plant Management and Organization

As has been stated many times, the vast majority of destructive fires start with carelessness in dealing with fundamentals—in other words, human carelessness. Correction of this situation requires, at the very beginning, interested management. Managerial interest alone, however, is not enough. To play any real role in preventing destructive fires or, for that matter, fires in general, management interest must be converted into effective action. In other words, management must develop appropriate policies and programs and create effective employee organizations to carry them out. Knowledgeable fire prevention personnel can be of great assistance in getting management interested and in helping management formulate sound policies and effective programs and create good supporting organizations.

I. Management. All well-managed businesses realize that safeguarding lives, property, and continuity of production is a necessity if they are to stay in business. They recognize the need to be as immune as possible to serious damage from the forces of nature and from fire. What is not so well recognized (the fire loss record proves this) is the extent to which self-help can be a primary tool in preventing fire losses. Too often, the entire load of preventing large fire losses is handed to fire departments. These organizations are primarily geared to fighting fires, providing technical assistance on request, presenting educational programs for the general public, and the like. Firefighting organizations do not have the time, manpower, or resources to develop detailed policies and programs for individual plants and other facilities, nor should they be expected to do so. These jobs should be done, and in many instances are done, by plant managements. Where policies and programs have not been developed, advising fire prevention personnel should encourage such action.

A. Policy. Top management should establish general policy. In addition to outlining the general policy to prevent fire losses, a policy statement should—

1. state the extent to which policy making is delegated to subordinate operating supervisors, defining both authority and responsibility (the boss can't be everywhere);
2. detail incentives for special work;
3. set priorities for accomplishments of objectives;
4. provide for cooperation with fire departments and other organizations; and
5. establish chains of authority and communication.

In smaller plants, it is quite possible that all policy making may be vested in one or two individuals, but the policy elements are still the same.

B. Program. Depending on the size of the plant and the nature of its operations, the program developed may be quite simple or extremely detailed. Some of the most important goals to be considered for inclusion in the program are listed below.

1. Scheduling improvements needed for greater fire safety in existing facilities.
2. Inspection, testing, and maintenance of fire detection and protection equipment.
3. Regular fire prevention inspections and correction of deficiencies found during inspections.
4. Organizing, educating, and training a plant fire organization.
5. Preplanning for proposed facilities.
6. Coordination with fire service organizations and with other organizations as appropriate. For example, a plant manufacturing, storing, and shipping hazardous chemicals might need to co-

ordinate fire planning with police departments, hospitals, and others.

7. Maintaining employee interest.

II. The organization. Each plant or facility should have a fire emergency organization. The size of such organizations ranges from one person to a full-scale fire department. Whatever the size, fire safety responsibilities must be lodged somewhere, or the whole program will be ineffective.

  A. Basic requirements.

    1. Familiarity with the plant or institution, the hazards involved, the fire protection systems, the utility controls (for example, gas valves), salvage procedures, exit facilities, and the like. Such familiarity is critically important for the solitary inspector or for the officers in charge of local firefighting forces.

    2. Education and training appropriate to the responsibilities involved.

    3. Suitable tie-ins with fire departments and other fire safety organizations.

    4. Knowledge of evacuation procedures.

  B. Recognition. Most persons involved in plant or institution fire safety programs carry these responsibilities over and above their normal job responsibilities. In many instances, educational and training activities must be conducted during overtime. It should be obvious that these extra efforts must be recognized in one way or another, or interest will soon lag.

Summation

General order and neatness (or housekeeping); emphasis on preventive maintenance; recognizing and safeguarding key operations; preplanning for new facilities, shutdowns, and disaster control; interested management; sound programming; and effective plant organizations form a set of tools that make for an excellent fire loss record. Accomplishing these goals does not require detailed knowledge of highly complicated special operations. It does require that people develop habits of logical thinking and a working knowledge of basic fire behavior and that they apply common sense. These attributes are readily within the reach of all fire prevention personnel. For special situations outside an individual's general and specialized knowledge, there are always sources of information that can and should be tapped.

chapter 14
Fire Loss Investigations

General Investigative Procedures

I. Goals. Two of the most useful tools for building a successful fire prevention program are thorough fire loss investigation and well-written, factual reports of findings. Through such investigations and reports it is possible to learn what is burning and why. By keeping and analyzing both individual and cumulative reports, it is possible to—

   A. come up with technically sound and economically feasible means to prevent a recurrence.
   B. spot trends up or down in specific areas and take appropriate preventive action in advance; and
   C. locate defective equipment and arrange for its replacement. For example, the author remembers three fires of varying magnitude that occurred over a span of a few weeks. All of the fires originated in the same place in a newly marketed brand of radio. Informed of this, the manufacturer rechecked the design and found that one small component tended to overheat. This defective component was replaced in units previously sold and in units being manufactured, and there were no more fires from this source.

II. Good and bad investigations. There are three ways to make a fire loss investigation.

   A. The windshield variety. The facts don't add up. How can an investigator give "careless smoking" as a fire cause when the plant had been shut down and secured four days before the fire? Or another example: Cause is listed as probably electrical short circuit. Sounds plausible until someone other than the investigator points out that electric power had been cut off several weeks earlier. Sorry, Mr. or Ms. Investigator—for all practical intents and purposes, you didn't get out of your car. Such slipshod work reflects unfavorably on the fire prevention organization and individual(s) responsible.

B. The doorway inspection. Here the investigator has gotten out of the car long enough to know that the plant was operational. The inside of the building is a shambles, so, after a quick look, and knowing that possibly 40 percent of all fires have electrical faults or careless smoking as a cause, the investigator picks one of them out of the air and sticks the word "probably" in front. If such investigators give the cause as probably electrical short circuit, they don't say where or why the fault occurred, nor do they give any clue as to why other potential sources of ignition were ruled out. Repeated second-rate investigations of this kind give the organization or individual responsible a reputation for generally poor work.

C. The complete inspection. Here the investigator does a complete job of checking out all possible sources of ignition, reasons for size of fire, effect on production, and other important facts. The report is factual, complete, and provable. Investigations of this type make the organization or individual look good.

III. Differences between arson and nonarson investigations. Investigating procedures for arson and nonarson fires are similar in many respects. There are, however, some differences that require comment.

A. Evidence in nonarson fires does not require the degree of security that arson evidence requires. Also, evidence for a nonarson fire may be discarded, if desired, as soon as it has served its purpose of determining cause and, in some cases, of pointing to appropriate remedial action.

B. Questioning regarding nonarson fires is on a much freer basis than that for arson fires because no crime is involved. There is also more leeway in report writing.

C. When the cause of a nonarson fire has been established beyond reasonable doubt, it is normally not necessary to take the time to rule out unlikely causes of minor importance.

Chapter 11 included some information on fire behavior in buildings; these comments, though pertinent, will not be repeated. Instead, emphasis will be given to methods of investigation that are common to all types of fires.

The Report Form

A good individual fire report can tell a great deal about conditions before and after a fire in one building or a single group of buildings. It can also point out appropriate corrective action for the buildings. By itself, however, it is useless in establishing broad policies or in writing fire codes. Effective action in these areas requires the analysis of the collective experience gleaned from a reasonable number of reports. For this to be done, it is necessary to have a uniform method of reporting. Narrative formats may differ, but the key items must always appear in the same place in the rating block. One hopes there will eventually be a nationally recognized form for reporting fires. NFPA Standard 901 is a major step in that direction. Pending the adoption of a national standard form, locally developed forms at a minimum should include the items shown in Figure 19.

```
Date
Time of Fire          Time Fire Extinguished
Location (community and state)
Name of Owner
Name of Occupant
Building Number or Numbers and Street Address
Point of Origin (section and/or story of building)
Occupancy
Method of Notification (box, fire phone, sprinkler alarm, etc.)
Discovered by (watchman, guard, passerby, etc.)
Building Information
    Kind of Building (warehouse, machine shop, office building, etc.)
    Number of Stories (include attics, basements, crawl spaces,
        as appropriate)
    Dimensions        Length          Width
    Wall Construction
    Floor Construction
    Roof Construction
    Roof Covering
    Interior Finish (note drop ceilings etc. as well as material)
Description of Damage
    To Building
    To Contents
    Other (vehicles, yard storage, etc.)
Automatic Sprinkler System
    Make of Head          No. Heads Operated
    Type of System (wet, dry, deluge, preaction)
    Supervised           Unsupervised
    Operation            Satisfactory          Unsatisfactory
    Protection Restored          Date          Time
```

Automatic Alarm System
    Name and Type of System
    Supervised   Unsupervised
    Operation   Satisfactory   Unsatisfactory
    Restored to Service   Date   Time
Equipment Used to Extinguish Fire
    Hose
    Extinguishers
Motorized Fire Apparatus Responding
    Local   (no. and types of trucks)
    Mutual Aid   (ditto)
No. Fire Department Personnel
Casualties
    Deaths
    Injuries
Weather (clear, snow, rain, etc.; wind velocity and direction)
Fire Cause
Factors Contributing to Extent of Damage (delayed alarm, weak water
    supply, etc.)
Previous Inspection
    Date
    By Whom
Amount of Loss
    Building
    Contents   Occupants of Area of Fire Origin
                        Other Occupants
    Other (vehicles, yard storage, etc.)
    Total

I. The rating block. Two comments on the rating block are necessary.
  A. In many fires, not all of the items indicated are applicable. For such items, the practice should be to insert the letters NA, meaning not applicable. This assures the reader who has to act on the report that nothing had been overlooked.
  B. It should also be remembered that an arson report is also a fire report. Information that may have no bearing on an arson prosecution can be essential in pinpointing structural and occupancy deficiencies that could be of importance in other locations and might be serious enough to warrant code changes.

II. The narrative report. In the narrative portion of the report, trace the sequence of events from fire origin to fire extinguishment in chronological order. Names and occupations of individuals directly involved may be included as appropriate. Keep the report as brief and clear as possible: nothing is gained by a long, rambling narrative.

    The narrative should not include data that are in any way inconsistent with the rating block. For example, if a rating block gives de-

Fire Loss Investigations 267

fective wiring or probably defective wiring as the cause, but the narrative says the cause may be of electrical origin but is still under investigation, the two reports are not consistent. Space in the body of the report permits a complete description, but it is normally necessary, in shortening the description to fit the rating block, to use care in the selection of words so the two will be in agreement.

When arson is suspected, use the terms possible or probable incendiarism or words to that effect. Fire reports are not privileged documents, and it is unwise to state or imply that a crime has been committed.

Some Investigation Techniques

Thorough fire loss investigations provide a great deal of the information needed by code writers, inspectors, fire departments, and others to lessen the hazards from fire. Such investigations are the result of careful training in determination of fire causes, why fires spread, and what caused the casualties.

I. Outside techniques. Many, probably the majority, of the serious fires occur during nonoperating periods, when no one or only a watchman is on the premises. For such fires, it is usually necessary to start the investigation outside the building. Failure to do so may lead to faulty conclusions which in turn result in unreliable fire reports. Various kinds of investigative techniques are outlined below.

A. Electrical. Major buildings as a rule have controls and other key electrical equipment—for example, switchgear, transformers, substations, and main switchboards—located in outside buildings, in cut off rooms, in vaults, or outdoors. It does not take long to find out whether or not power was on or off during the fire. Neither does it take long to find out whether or not there had been any unusual electrical disturbance, such as lightning, on the lines.

1. If the power was definitely off, this knowledge will save a lot of time inside the building.

2. If there was an electrical storm at the time of the fire, a lightning check is definitely in order. Lightning can and has entered buildings through supposedly dead lines. Check points of wiring entry into buildings for evidence.

3. As a logical extension of the outside check, main internal controls should be checked first for condition of switches, fuses, and breakers. If there is reason to suspect a wiring fault and breakers have not tripped, breakers should be checked out by a qualified testing laboratory.

B. Flammable liquids and gases. It is quite common to find small to moderate amounts of flammable liquids in drums and flammable gases compressed in cylinders stored close to a building. Even more common are small- to moderate-sized LPG-dispensing containers. Where such storage and/or dispensing operations are known to exist, the presence or absence of windows, doors, and other types of wall openings should be checked. Of particular importance is whether or not there were intake fans, especially fans connected to air handling systems. More than one fire has started when flammable vapors or gases from leaking containers entered a building and met an ignition source.

C. Exposure. While it is fairly easy to pick up major exposures (it would be difficult to miss such exposures as a lumberyard, a paint and varnish plant, or a wood frame building), the possibility of small exposures is often overlooked. It must be remembered that, where wall openings exist, a small external fire can enter the building and start a major fire inside. It is particularly easy for fire to go through small openings between panels or at the floor lines of metal buildings. Usually such fires are the result of the ignition of relatively small amounts of trash (in or out of containers) that has been allowed to accumulate too close to a building. Less frequent, but still fairly common, is fire in windblown debris that has accumulated under a building because there were no walls or screening to prevent such accumulations. In most instances where trash fires have started building fires, enough evidence is left to prove that the building fire started at the wall and that there was a trash fire outside. The writer has investigated a number of such fires, the largest of which caused a loss of about $400,000.

D. Summation. In the preceding paragraphs, three major outside ignition potentials have been outlined. Other potentials may also be present. None of these items should be ignored: they should be checked out. Failure to do so may mean that the conclusions drawn from an inside-only investigation are faulty.

II. Inside techniques. Before new fire prevention personnel are assigned to investigate major fires, they should have become reasonably familiar with fire behavior, the more common causes of ignition, and basic structural, occupancy, and protection defects that may play a major part in converting a small fire to a major one. Guidelines on these subjects were furnished in preceding chapters. Where possible, training programs should include provision for new personnel to accompany experienced investigators to a reasonable number of fires.

Fire Loss Investigations 269

Finally, the investigator needs to clearly understand that the purpose of the investigation is to be able to answer the following questions. (1) Where did the fire start? (2) What was the ignition source? (3) Why did a small fire become large? (4) What were the reasons for casualties (if there were any)? (5) What was the amount of loss? (6) Why did the fire pattern deviate from normal (if it did)? (7) Was the fire likely to cause indirect losses away from the fire scene?

There are normally two lines of investigation following a major fire: (1) tracing the fire, and (2) determining the loss. Both lines can be followed simultaneously, but this chapter will only outline methods of tracing the fire.

A. Point of origin.

1. In the initial check, temporarily disregard the upper portions of the fire, and focus attention on locating low burns.

2. For each low burn, note location of main fire by depth of char and other indications, and analyze the spread of fire away from this point.

3. Into the basic pattern fit in wind direction, chimney effect of floor openings and shafts, and similar influences on spread of fire.

4. Although there may be a number of low burn points, normally only one will appear to be the logical place for the fire to start. Minor low burns quite often can clearly be identified as resulting from the developing fire and not starting it. If, however, two or more low burns appear to be of nearly equal importance to the developing fire, the odds are ten to one the fire was set (at which point an arson investigation is in order).

5. Assuming the major fire originated at a single low burn point, the next procedure is to determine the cause.

B. Source of ignition. Ignition sources may be open flames, sparks, steam pipes, hot oil, motor bearings, or any one of a number of other items. For these sources to be responsible for serious fires, however, it is necessary for some fuel to be present. The kind and amount of fuel present can be quite important in determining the ignition source. For example, a carelessly discarded burning cigarette will not ignite a wood floor, a wood box, or a tight roll of paper. It will ignite loose paper and many other materials. Such organic oils as linseed oil and most inorganic oils, including mineral oil, burn—but only the organic oils spontaneously heat.

There are far too many potential ignition situations to detail in any one book, but a working knowledge of basic principles as illustrated in the following examples will usually lead to the right answer.

1. Electrical.
    a. Following a major fire, electrical wiring, conduit, fixtures, equipment, and appliances are scattered about. In general, the wiring proper offers no clues except that of tracing a circuit—which can be important. Nevertheless, there are usually some other clues that can be checked out. One such clue is the fusing of the ends of copper wire at a breaker or another control point.
        1. A simple heavy fusion at a control point, accompanied by a blown fuse or a tripped breaker, is substantial evidence that a short circuit occurred and that the safety equipment was activated. The next point to check is whether or not fuel was present that could have been ignited by the short circuit. To answer this question, an inspector needs to know what the prefire occupancy was and where the circuit was.
        2. Multiple fusion points generally indicate that the fire source has not been located. It is probable that there were short circuits along the line but that the breaker or fuse failed to shut off the circuit on the initial short circuits. It is possible that the breaker did not trip at all. This does not necessarily mean that the fire was not electrical in origin. It does mean that it is quite possible that the short circuits might well have been caused by an external fire, and this should be checked out.
    b. Although possible, it is extremely rare for fire to be caused by a short circuit in conduit. Some conduit, however, is galvanized, and zinc has a low melting point. Heat from an external fire can melt the zinc and cause a short circuit inside the conduit. Such a short is the result of fire and not a fire cause.
    c. All electrical equipment is vulnerable to single phasing. Causes of single phasing on the usual three phase system include a broken utility line, contact failure, wrong lead connection (new work or recent rewiring), a grounded conductor, and other problems. Under such circumstances, equipment that has not been protected to compensate for single phasing, overheats and becomes a potential fire source.
    d. Many fires of electrical origin are caused by the use of improper or incomplete equipment. Quite often there is enough evidence left to check this out, particularly in fire resistive buildings where much of the structure is left standing. Two simple illustrations follow.
        1. If lightning is suspected, check the existence and adequacy

Fire Loss Investigations

of lightning protection. If protection is missing and lightning did strike, the lightning's point of entry into the building can be plainly seen, as can part of its path inside.

2. Pendent lights on drop cords are widely used in place of the far safer pipe pendent lights. Where combustibles are present, a light bulb on a drop cord should have a wire guard and, in hazardous areas, a vapor-tight globe as well. There will be evidence of the guards and vapor-tight globes, if they were present. Lack of a wire guard cost a government agency over $7 million a few years ago when a burning light bulb in contact with combustibles was not turned off.

2. Flammable liquids.

a. Strictly speaking, flammable liquids do not cause fires—but many serious fires have resulted from their misuse, from faulty design, and from equipment failures. Among the most common sources of serious flammable liquids fires are the following.

1. Failure to control ignition sources, in many instances, is more important from a life safety standpoint than from a property standpoint. For example, a small group of men was cleaning paint spots off the concrete floor of a newly painted building. After running out of the safety solvent, they siphoned gasoline from a car to complete the job. Gasoline vapors were ignited by the pilot light on a water heater. In the ensuing flash fire, four men died. Damage to the building was ninety-five dollars. Similar fires have, of course, caused substantial property loss. Where misuse of flammable liquids is suspected, it is important to make sure that possible ignition sources for the vapors are not ignored. This is particularly important when vapors could have penetrated a lower story or a basement.

2. Quench oil tanks are at times the source of serious fires. Usually, such fires result when the oil is overheated for one reason or another or when hot metal articles get hung up on the passage through.

3. Poorly arranged flammable liquids piping has contributed to serious fires. The writer knows of two instances where a fitting failed and flammable hydraulic oil sprayed directly onto furnaces. Each incident resulted in losses of over a million dollars.

b. Where the involvement of flammable liquids is suspected, an inspector can quickly check for their presence by placing fire residue in a closed container of water. Most flammable liquids likely to be involved in such a situation are not soluble in water and float; a certain amount of color at the top of the water indicates their presence.

c. Investigators commonly find containers at a fire scene. A container's type and condition and the presence or absence of flammables can tell a great deal. For example, it is reasonable to assume that a safety can that has bulged and split at the seams may have contributed to the fire but was not involved in the beginning. A safety can or other sealed container that leaked would be unlikely to build up enough internal pressure to split a seam.

d. Heating equipment. Many potential sources of trouble with heating systems have been explored in previous chapters. This section comments only on some methods used to uncover the fire source.

1. Check steam pipes. A three story, unsprinklered, wood frame building was involved in a serious fire while unattended. Prompt fire department response held the damage to one wing. A preliminary check indicated no reason to suspect electrical faults or misuse of flammables. It was noted, however, that the building was heated by wall radiators with piping between stories concealed. Random removal of screening showed that the clearance between the steampipes and the wood flooring ranged from $\frac{1}{8}$ in. to zero. At points of contact with the steam piping, the flooring was charred. If this situation had not been picked up and corrected, another fire would have been inevitable.

2. Check vent pipes. A multisection building involved in a fire had four-hour walls between sections. At one time, part of the end section had been heated by a solid-fuel stove. The vent pipe for the stove went through the wall into the next section and then vertically up through the roof. None of the other sections was heated, and all were used for warehousing. A fire originating near the stove vent pipe virtually gutted the adjacent section. Investigation revealed the following facts: (1) since the end section had not been heated in a long time, the tenant in the adjacent section stored combustible merchandise in close proximity to the vent pipe; and (2) the tenant in the end section decided to dispose of a large amount of combustible debris by burning it in the stove.

Heat from the fire was transmitted by the metal stack and ignited the contents of the adjacent section.

(3). Check dial records. It is a common, although not universal practice at many heating and power plants to automatically record various dial readings, including those indicating amounts of flammable gases in exhaust stacks, boiler water level, and pressure. In many instances, these dial records survive a boiler explosion and any ensuing fire. These records should be salvaged. They may not mean too much to the average investigator, but they can tell a great deal to a boiler expert.

e. Miscellaneous sources of ignition. As has been previously noted, there are thousands of potential sources of ignition. Some are quite obvious to a trained inspector, and many can be determined by the evidence of eyewitnesses. There are, however, some potential sources that are frequently missed because they are outside the usual sphere of experience. For example:

1. When a fire appears to have originated close to high-speed machinery, consideration must be given to the possibility of friction as an ignition source. Underlubricated bearings on high-speed equipment can generate enough heat to ignite lint, flammable liquids, paper, and various other materials. When friction is suspected, it is worthwhile to check bearings of machines not heavily involved in the fire as well as past inspection reports to determine general maintenance of high-speed machinery.

2. Gas confined in piping is not a hazard. Gas may leak from a pipe because of failure of pressure joints, line breaks or fractures, poorly sealed connections, poor valves, or other defects. If there is no source of ignition at the leak and the leak goes undetected until a gas-air mixture in the explosive range has formed, there will be an explosion ahead of the fire. In contrast, failure of a gas line from heat during a fire results in a blowtorch at the opening and does not generate an explosion.

3. On some occasions, severe fires have resulted from cutting and welding that was not being done on the fire floor. For example, an air handling duct for a factory and warehouse was being remodelled. Adequate precautions had not been taken to prevent globules of molten metal from leaving the cutting and welding area. One of these molten globules

passed undetected via the ductwork into another story. There, in the absence of human occupants and sprinkler protection, the hot metal ignited combustible contents. Damage was over $1 million.

4. Assume that a file room with rows of both metal and wood files has had a major fire. A preliminary check shows that the bulk of the damage was from 3 ft. above the floor to the ceiling. The preliminary check also indicates that the fire originated near two or three adjacent tiers of cabinets. When it is clear that the fire did not originate at the floor level, one of the first things to be done is to open the file drawers at the point of origin. In many cases, it is possible to determine the drawer in which the fire started by comparing the amount of fire damage to the several drawers. It is then relatively easy to find out what was in the drawer prior to the fire and check out the possibilities of fire origin—for example, spontaneous heating, unauthorized materials, or careless smoking.

f. Summation. While it is true that, in the majority of cases, fire causes can be fairly quickly determined from eyewitness accounts and obvious clues, there inevitably will be a number of fires for which there are no eyewitnesses and no obvious clues. In such instances, the investigator should learn to ask himself or herself questions based on the observed fire pattern and basic knowledge of fire behavior. For example, knowledge of gas characteristics can definitely pin down whether gas was the causative agent or merely a contributing factor to a fire initiated elsewhere.

C. Reason for a small fire to become large. Equally as important as determining how and where fires started is finding out why a small fire became large. The reasons behind major fires need to be discovered, not only to assist in preventing future major fires of the same type in other locations, but also because most major fires also involve indirect losses at locations remote from the fire.

1. Siting. A few years ago, virtually all major industrial and commercial buildings and most large institutional facilities were located in heavily built-up areas. In almost every case, a well-staffed fire station was located nearby. Today, many of these large-scale operations are located well away from metropolitan areas. Unless these organizations are large enough to have their own fire departments, they are also likely to be far from a well-equipped and

-staffed fire fighting force. Response time can be critical in the event of a sudden emergency situation. For example, assume a plant with a 20,000-gallon gravity tank and an 8-in. connection to a 10-in. public main. In a city, a 10-in. main is part of a grid, and there is a good hydrant system. In a rural area, a 10-in. main is generally ungridded and may run several miles to the water supply. Assume a flash fire and fifty heads open. If each head draws 20 gpm, the sprinklers alone require 1,000 gpm, and the hose streams also need water. If the public water connection is out of service, the tank will be empty in perhaps fifteen minutes. As a result, fire response time is critical.

    a. Metropolitan area. Almost always, the fire department response time is less than five minutes. A hook-up would quickly be made to other mains, and the sprinklers would stay in service with adequate water. Except under unusual circumstances, the fire would be controlled with only moderate losses.

    b. Rural area. Due to distance, weather, and other factors, response time is over twenty minutes. The nearest water supply is 2 mi. away. By the time the necessary hose relays have been put together, the sprinklers have exhausted the water supply. The result: a major fire for want of a few minutes.

Definitely, siting can be a major contributing factor to fire spread.

2. Delayed discovery. Far too many buildings of importance do not have watchmen, fire detection systems, or fire protection. Some others have local detection systems but no other safeguards. Churches, schools, many warehouses, and various other types of buildings are examples of this lack. Many of these structures are unattended nights and during a substantial number of daylight hours. Under such conditions, most fires are detected by some casual passerby. By that time, the fire has generally developed to the point that a major loss is inevitable.

3. Construction and occupancy. The part that construction and occupancy defects might play in a major fire was discussed at some length in earlier chapters, and these comments will not be repeated here.

4. Delayed alarms. All too often, employees voluntarily or by instruction attempt to fight fires with the means at hand prior to calling the fire department. Unless the fire is so small that there is no question that it can be extinguished immediately (e.g. with *one* extinguisher), the first firefighting priority is to call the fire

department. More than one life has been lost and many millions of dollars worth of property have been destroyed because untrained or partly trained employees took a chance on their own.

5. Santa Ana winds. In the Southwest, a great deal of property and some lives have been lost because of a stubborn refusal to recognize what wind can do in dry country. Santa Ana winds are hot, dry winds that come out of the desert and may attain velocities exceeding 100 mph. Despite the frequent occurrence of these winds, many people take a chance and plant combustible shrubbery close to their basically wood homes and then use wood shingles or shakes for roofs. The result is inevitable: a fire starts somewhere, and gale-force winds sweep the fire through dried out woods and these "rustic" houses. It should be noted that military installations, many industrial plants, and many houses have survived such conflagrations with no or very little damage because of proper arrangement of buildings. Weather can be a major factor in the size of loss.

6. Water supply. To some extent, water supply has been previously discussed. A hazard that has not been mentioned is the possibility that water lines and hydrants may freeze. Usually such freeze-ups occur irregularly, but they can result in delay in getting water to sprinklers and/or hose lines.

7. Summation. Fire inspectors generally recognize such construction defects as inadequate cutoffs and too large areas; such protection defects as weak water supply and lack of sprinklers; occupancy defects; and exposure problems. However, other factors may also help a small fire become large. Some of these other factors are noted above. Such factors should not be overlooked, as the part they played in the overall picture can be very important information to a number of responsible officials.

D. Reasons for casualties. In the United States alone, approximately 12,000 people die annually as a result of fire. This toll does not include deaths, supposedly from heart attacks, of persons in areas immediately above or adjacent to the fire. Contrary to popular belief, the majority of fire deaths are not from burns. (The fact that a recovered body has 100 percent third degree burns does not mean the individual died from burns.) Careful autopsies have proved many times that death occurred prior to burning. Causes of such deaths include the following:

1. Carbon monoxide poisoning.
2. Carbon dioxide suffocation.

Fire Loss Investigations 277

    3. Oxygen depletion.
    4. Excessive temperature.
    5. Toxic gases from burning materials.
    6. Smoke inhalation.
    7. Murder.
    8. Suicide.

While a fire investigator is not responsible for investigating fire deaths, he or she should make certain that an autopsy to determine the cause of death is made. (Insurance investigators may not get involved in this phase of an investigation, but public investigators should.) Autopsies are normally performed by a coroner, unless arson is suspected. In arson cases, the cause of death should be checked by a specialist in forensic medicine.

Information obtained from autopsies can be of great value in designing safer buildings and products.

E. Amounts of loss, direct and indirect. After a fire has been extinguished, it becomes important to arrive at a fair estimate of the actual damage incurred. The estimates arrived at affect insurance carriers who pay the losses and fire departments whose budget requirements may be influenced. Fire losses also hurt the general public through loss of employment and the like, and various governmental and private agencies help pick up the bill for unemployment and injuries. To arrive at a true loss picture, it is necessary to take many things into account.

    1. Direct property loss.

       a. Insurance company loss payments are based on the extent of their liability and not necessarily on the total loss. An underinsured company becomes a coinsurer to the extent that it is underinsured. For example, if a $100,000 fire occurs in a plant that should have had $1 million in coverage but actually has half of that, the insurer is only responsible for half the loss.

       b. Most initial insurance company reports on building fires include estimates for both insured and uninsured losses. There is normally a breakdown among losses to building, contents, and outside storage. These reports, however, generally do not list damage to such property as privately owned vehicles on public streets.

       c. Initial reports are not necessarily the same as adjusters' reports. Adjusters' reports take into account such items as coinsurance, building and equipment repair or replacement costs, stock inventories, and depreciation.

d. An often overlooked item in losses is the possibility of salvage. If plants take prompt action to dry out wet items and clean equipment, losses can be considerably reduced. Similarly, insurance companies frequently work with salvage companies to reprocess and salvage goods for resale.

e. Although fire department officers generally do not get involved in investigating and adjusting losses, they should have a working knowledge of how to judge the condition and maintenance of the building involved in a fire. This judgment, together with records from building and assessors department and an investigation immediately after the fire, may be the only way to get a realistic loss estimate for an uninsured property. In some instances, it may also be desirable to have an outside appraisal made.

f. It is necessary to make a distinction between self-insured and uninsured. A number of large corporations with many plants are self-insured. They have insurance departments that are, in effect, insurance companies, and these departments are funded to take care of possible fire losses. Uninsured means simply that the company or individual has made no provision for covering fire losses.

g. Summation. To make a reasonably accurate public fire loss estimate, it is essential that all the factors be known. A giant step in this direction is to establish good working relationships with insurance carriers, adjusters, building departments, assessors, and similar sources of information. Collectively, the information from these sources provides excellent fire loss data.

2. Indirect fire losses.

a. An important part of the losses from many fires, even some small ones, is the effect on production or operations. Many businesses today carry insurance for production losses from fires. When this is done, insurance payments provide a reliable estimate on the loss to the company involved. For noninsured organizations, however, an estimate can and should be made. For example, an inspector should determine such factors as approximately how long an operation will be down and capability to transfer operations to another location.

b. Any major fire involves losses to others than those directly involved. At present, there is no practical way to estimate this "ripple" effect, which may take the form of shutdowns at other locations, welfare increases, and other costs. Even though these cannot be estimated, their potential cannot be ignored. The degree to which these factors may be present can affect overall fire prevention decisions.

Summation

Much useful information is obtained from well-written reports of thorough fire loss investigations made with a good understanding of the factors involved. Findings from such reports are used to determine, not only fire losses, but also corrective actions, guidance for other similar locations, relative effectiveness of various types of extinguishing agents, salvage methods, available ways to cut production losses, possible needs for code revision, and the like. In the past, many improvements have been made as a result of good investigations, and, as the caliber of investigations and reports continues to improve, the future should show even more improvements in the fire loss picture.

chapter 15
Public Relations and Education

Careful analysis of the fire loss record clearly shows that at least 95 percent of all fires are caused by people, mainly through ignorance of fundamentals or through carelessness. With few exceptions, individuals do not by choice suffer financial loss or personal tragedy from fire, nor do they wish such ill fortune on others. People will cooperate in reducing the fire risks if they are shown how they can help. Good public relations and sound educational programs are the keys to success in this area.

It is unfortunately true that, in general, public relations and fire prevention education programs are the weakest links in overall fire prevention programs. In far too many communities, fire prevention public relations programs are limited to some special events during Fire Prevention Week, possibly a more modest program for Spring Cleanup Week, and an occasional small article in the local press. In these same communities, educational programs are largely limited to infrequent talks at elementary schools and various amounts (usually quite limited) of directions and instructions at places of employment. Except for such limited programs, public contact with fire prevention in many communities consists only of an infrequent, fleeting contact with an inspector visiting the place of employment or the sight of a uniformed fireman at a place of public assembly. Under such conditions, it is perfectly natural for the public to assume that the experts have taken care of fire prevention requirements and that there is nothing that they themselves can do. This fallacy is one of the big reasons why the United States trails the field among the industrial nations insofar as fire loss experience is concerned. Trained fire safety people can only do so much. Beyond that limit, the public must get involved if we are to accomplish the reductions in fire loss to life and property that are definitely possible.

Good fire prevention programs in the fields of public relations and education cannot be set up on an intermittent basis. They must function on a year-round basis and be arranged to appeal to the audience at which they are aimed. For example, a talk delivered to a group of architects for buildings would quite properly deal with technical problems of fire safety design that would be meaningless to a group of housewives. These women, however, might be deeply interested in life safety in the home. Only by year-round programs and by a readiness to deal with audiences and individuals on the basis of their fire problems and their interests is it possible to get the public involved and to keep them involved. Once this is achieved, fire prevention personnel have gained invaluable allies in the fight against fire loss. The proof of this may be seen in the loss records of organizations that conduct year-round public relations and educational programs. Governmental agencies and private concerns that do this with thoughtful consideration of the needs and interests of the employees concerned regularly produce outstanding fire loss records. It should also be noted that employees of such organizations carry the fire safety message home, which in turn has meant greater safety for their families.

Public Relations

I. The media. One of the most effective ways to reach the public is through such mass media as newspapers, magazines, television, radio, billboards, and flyers. These sources have perhaps not been used as effectively as they might have been because many fire prevention personnel have limited contacts with the media and fear committing limited funds to advertising. These deficiencies can be overcome, however, by giving consideration to the situation in the area in which the fire prevention organization is working and then taking appropriate actions. For example, fire prevention personnel should consider the following suggestions.

    A. Develop and prepare material in a style that is interesting and well written. Vague and poorly written material can and normally does land in a "File 13" trash can.

    B. Prepare material in a format that is suitable for the audience at which it is aimed.

C. Don't bombard editors with a daily mishmash of platitudes and vague technical details. Well-written articles at intervals receive attention that "junk" mail doesn't.

D. As appropriate, cultivate a working relationship with editors and the working press. These relationships pay good dividends for both parties. For example, reporters may get help with technical details of fires or codes, and fire prevention personnel may gain editorial coverage or page-one space from time to time.

E. Magazines such as the *Reader's Digest* are always on the lookout for well-written articles on almost any subject that can be named. Getting an article in such a publication means reaching an audience of millions.

F. Radio broadcasting lends itself admirably to talk shows and question and answer shows. Many radio stations, as a public service, are quite willing to put on fire safety talks for free. All that is required is the ability and willingness to talk with competence and with an easy manner to the radio audience. Develop a stage presence, and don't worry about the microphone. It won't bite.

G. Long television programs do cost money that is not available to the average fire prevention organization. A well-heeled organization may set up a fire safety show of this magnitude, but the average fire prevention organization must take a little different approach. The two primary ways of getting local TV coverage free are described below.

1. Provide a station with short film clips (thirty seconds). Many excellent film clips are available, or a fire prevention organization could create one of local interest. TV stations show these in the public interest.

2. Be available for interviews on the local news. Do not get involved in such an interview without being prepared for it.

II. Potential supporting agencies. Nearly every community has one or more organizations with a vested interest in limiting fires. Chambers of commerce, boards of trade, various manufacturing groups, business clubs, and others have a real interest in preventing fire losses. Get acquainted with them and, where desirable, be prepared to give short talks on phases of fire prevention that are of direct interest to the group. Such contacts pay off in two ways.

A. Group members carry fire safety ideas back into their businesses.

B. Frequently, groups or members underwrite the cost of paid advertisements, provided the ads are well thought out and well written.

III. Lectures and demonstrations.

A. It is a relatively simple matter to make up a small demonstration to drive home a few points of vital importance. For example, all that is required to demonstrate that flammable liquid vapors are heavier than air is a simple wood and metal stand, a piece of glass or plastic tubing open at both ends, a base for a candle at the lower end of the tube, and a small container attached to or adjacent to the top of the tube. To use, place a small amount of low-flash point solvent in the upper receptacle; the vapor will flow down the tube, become ignited by the candle, and flash back up the tube. More than one person has died or been seriously burned by using flammable liquids for cleaning upstairs while a furnace was running downstairs. Similar inexpensive test setups can be made to illustrate common hazards.

B. It is helpful to be able to present lectures to various groups. To do this requires the development of stage presence so the presentation will not be stiff or awkward. Slides, if available, are usually very helpful.

IV. Special events. Over and above year-round activities, it is desirable to highlight fire prevention efforts with special events such as Fire Prevention Week. While it is in order to have parades, Miss Flame selections, dinners, and so on, these should all be tied together to put the spotlight on fire prevention. Too many communities forget that and wind up with an affair that is a nice holiday but does nothing for fire prevention.

V. Summation. The public relations methods cited above are only some of the ways in which fire prevention can be kept in the public eye and, over the long run, gain public cooperation. Not all the methods should necessarily be used at any one location or by all fire protection personnel. Selection of methods should be based on the needs of the community and the best approaches to the various types of people living and working there. Whatever methods are used, it is important that the listener or reader get the impression that the fire safety personnel involved are not only competent but are also interested in the public's problems as a whole, in groups or as individuals. Neither competence nor personal interest can be faked for long. If an inspector is queried on a subject outside his regular field, it is far better to say "I don't know, but I'll find out for you," than it is to guess.

Education

It has been the writer's experience that the number of ways to mount effective fire prevention educational programs is virtually limitless. Efforts do not have to be confined to elementary schools, valuable as such programs are. Effective programs have been set up in the form of seminars on topics of particular interest, training sessions for plants and institutions, special talks to clubs or youth organizations, and in many other forms.

I. General principles. Some of the programs observed by the author have been extremely good. Others have failed completely. How do you educate? The first key to a successful educational program is to know your audience. Know their interests, the types of hazards they face, and the level of presentation required to hold their interest. Following are some comments on dealing with various kinds of audiences.

A. To deal with adult audiences, it is necessary to have a good stage presence, a reasonably good speaking voice, a topic suitable to the audience, and a readiness to answer questions to the best of one's ability. Language used should fit the audience. Technical terms are fine in dealing with engineers, but these terms must be converted to layman's language for talks to nontechnical people or, for that matter, to technical people in nonengineering professions. There is a tendency to forget that each profession and each craft uses terms that are peculiar to it and are therefore unintelligible elsewhere.

Don't be long-winded. That is about the easiest way known to lose an audience.

Where available, good short films, slides, and simple, safe tests are desirable additions to a lecture.

B. Presentations to children should be handled by someone who honestly likes youngsters and has a knack for knowing what will interest them. Keep the presentation short and in a language the children understand. When this is done properly, it is amazing how much youngsters pick up.

C. Know the difference between a voluntary and a captive audience. Voluntary audiences are present because they are interested in the subject and want to learn more about it. It is not necessary to capture their interest, but it is necessary to keep it. Captive audiences are not there because they want to be but because they are required to be. It is first necessary to get their attention, then to gain their interest, and after that to hold their interest. Obviously two different methods of presentation are necessary.

D. Keep up with the shape of things to come, not so as to become an immediate expert, but in order not to get caught flat-footed by a surprise question. The future holds such innovations as superairports, 500,000-ton tankers, offshore ports, and 300-mph trains. A questioner on this sort of thing will almost always be satisfied with an answer such as, "I'm aware of the problem you mention. I don't have all the answers yet, but will find out for you. In the meantime ..." (Add whatever is currently known.)

E. Finally a lecturer should practice what he preaches. No one will pay attention to what is said if it isn't practiced.

II. Schools. Well-designed fire prevention educational programs aimed at the young can be very productive, for such programs can lead to lifelong habits of fire safety consciousness. School programs start with the establishment of good working relationships with school officials. Once this has been accomplished, it becomes possible to draw up a schedule that will get the fire prevention message across without interfering with regular class work.

Because schools have different kinds of programs and bodies, it is unwise to have rigid programming. At one school, presentations at general assemblies might be the best approach; at another, visits to individual classrooms might be indicated. In the older age groups, student organizations concerned with civics or related matters might be the ideal audience. Advisory sessions after school hours may be desirable. Tailor the program to fit the school. Last but not least, select the individual (s) to carry out the program. On their shoulders rests the final responsibility for successful communication.

III. The home. Fire in the home is responsible, not only for a large part of the national property loss, but also for perhaps 40 percent of all fire deaths and injuries. Most of the people involved in these fires have little contact with fire prevention and little realization of the way hazards can pile up in a home. True, they may occasionally hear a little at work or from children coming home from school with what they have heard and perhaps with home inspection blanks. But those contacts in many communities are of limited effectiveness for one simple reason: There is no effective program for direct contact between fire prevention personnel and home owners or tenants. When such contacts are instituted, residential property losses and fire deaths take a sharp drop. This is a fact of life that should be brought to the attention of public officials and the press.

The first step in home fire prevention education is a public relations campaign to convince people that there is a genuine concern with their fire safety and a desire to help them improve their individual situations. Once this is done, it is fairly easy to secure voluntary compliance with home inspections.

In making home inspections, it is essential to remember that the operation is voluntary. Its primary purpose is to educate people about such hazards as rubbish accumulations, flammable liquids wrongly used, and overloaded electrical circuits. This can be done in most instances during inspections of attics and cellars. Information should always be imparted without any appearance of snooping or prying. For example, an inspector may find that the fuse box in the cellar or basement is a 60-amp unit, yet in plain sight are various power tools and other electrical appliances. The potential for an electrical fire is obvious and can easily be explained to the occupant. Personnel making home inspections should be aware of sources of help for the occupant. (For example, will city provide for rubbish removal.) Once a satisfactory home inspection program has been launched, the impact of fire prevention messages from other sources is much greater.

IV. Non-English-speaking groups. In some areas, particularly in the Southwest and in the larger cities, there are large numbers of people who do not speak English. Most of these individuals are fairly recent immigrants, and many of them work in factories, institutions, or commercial facilities and live in closely spaced, older housing. Lacking English, they are unwittingly responsible for fires out of proportion to their numbers. They have difficulty in understanding fire safety instructions and the uses of fire protection equipment, and the obvious result is more and larger fires than there should be.

Where non-English-speaking people congregate, the fire prevention bureau should make special efforts to see that its programs are made understandable to them. There is little problem in arranging for fire warnings, work hazard notifications, and instructions for the use of extinguishers to be translated into the appropriate language. This job can often be done by a bilingual member of the bureau. Such individuals are also valuable in maintaining community relations with these groups. Last but not least, bilingual instructors who handle the situation properly are actually giving a course in basic English.

V. Youth groups. It is possible to mount effective fire prevention educational programs through such youth groups as Boy Scouts and Girl Scouts, 4H clubs, and junior fire departments. Education among these groups is indirect as a rule and includes a wide range of activities. For example, a fire prevention officer may serve as a merit badge counsellor, help organize youth cleanup drives, and provide more specific fire prevention programs.

VI. Institutions. Special programs are needed for such institutions as hospitals, nursing homes, schools for the handicapped, and penal institutions. Although these institutions differ radically, they all have one thing in common: A sizable number of institutional occupants, for one reason or another, are unable in the event of fire to take action to assure their own safety. Accordingly, operators and employees of such institutions need education in eliminating unnecessary fire hazards, selecting the best available means to provide protection, and safely evacuating those who cannot take action on their own in the event of fire. The lack of such programs shows up in repeated headlines: "_____ people died in fire in _____, unable to escape."

There is no pat solution to the problem, since no two institutions are alike. To develop a program, it is first necessary to get a working knowledge of the institution's construction, occupancy layout, existing fire protection, exit facilities, and other pertinent information. From these data, it is possible to draw up a good program. In so doing, it is necessary to not lose sight of the fact that night occupancy may differ from day occupancy in very important ways. For example, investigations of a number of nursing home fires have revealed that staffing was excellent during the daytime, but very poor at night. Always keep in mind what the institution is set up to do.

VII. Seminars. A very useful tool in spreading fire prevention education is the seminar. Topics selected for lecture and discussion may range from a general overview of fire prevention to such specific occupancy problems as textile plant fire hazards, installed protection systems, or any other pertinent subject. Effective seminars must meet specific requirements in order to be effective.

    A. Invitations to attend should be issued only to those groups that have or should have an interest in the subject presented.
    B. Speakers should be knowledgeable and able to think on their feet.
    C. Question-and-answer sessions should follow lectures.
    D. Demonstrations should be provided as appropriate.
    E. Provision should be made for informal discussions where practical.

VIII. Civic Organizations. Many civic organizations have a direct interest in projects that benefit the community. Generally, such organizations are set up to include at least one representative from each important organization or group in the community. Fire prevention personnel should develop good relationships with such groups. By so doing, the personnel can arrange for appearances at organization meetings to talk about fire prevention. When such talks are effectively presented and cover subjects of interest to the group, the information generally spreads far beyond the immediate audience and benefits the various groups represented.

IX. Industry. Fire prevention in industry is usually considered to consist of fire brigade training, but it need not be limited to that. Employees in many instances need to be instructed in how to deal with fire hazards where they work. Many of the larger corporations recognize this. They have developed fire safety programs tailored to their operations and provided fire prevention sections to carry out the programs. Unfortunately, similar programs are not so common among small- to medium-sized concerns. Primarily, this lack is due, not to lack of interest, but rather to the inability to fund and/or man such programs. By acting as instructors or advisors, fire prevention personnel can help such companies set up part-time programs to meet the need.

Summation

The intent of this chapter has been to sketch a broad outline of some of the ways in which public relations and educational programs can be developed to aid in the goal of fire prevention. Not all of the methods noted are applicable everywhere, and, in some communities, other methods will suggest themselves.

For those developing public relations and educational programs, a word of caution is necessary. Neither fire prevention organizations nor individuals should spread themselves too thin: to do so would mean ineffective work in all areas. It is necessary to recognize whatever limitations on manpower and money are present and to establish priorities—that is, to separate the essential from the postponable and the nice to have. Effective work in areas for which manpower and/or funds are available will lead to grants for additional personnel and supplies.

chapter 16
Fire Prevention Inspections

Of all the tools available to make a real dent in the annual toll of fire deaths and injuries, property losses, business failures, community disruptions, and the like, fire prevention inspections are potentially the most useful. To accomplish the desired end, however, it is necessary that inspections become more effective and more widely used than is now the case in many communities. Effective inspections require a knowledge both of the objectives and of how to go about achieving these goals.

Goals

All fire prevention inspection programs should have as their primary objectives the following goals.

I. The elimination of fire causes where possible. (The fire that doesn't start can't burn anyone or damage any property.)

II. Minimizing the possibility of small fires becoming large. (Good fire cutoffs, fuel control, protection, and other precautions help promote this goal.)

III. Prevention of deaths and injuries. (Check exitways and, in some areas, recommend use of more fire retardant materials, or other safeguards.)

IV. Minimizing the impact of fire on community business and social life. (Stress safety for key activities.)

V. Provision of safeguards for inherent hazards that cannot be eliminated.

VI. Finding ways to accomplish objectives at reasonable cost. No recommendation for improvement should cost more than the loss it is designed to prevent. The converse is also true. For every worthwhile human activity, there is an economically sound solution to the problem of protecting against its fire hazards.

Attitudes

Much has been written elsewhere about the need for good grooming, courtesy, and other amenities on the part of fire prevention personnel. Accordingly, these obvious requirements will not be further commented on here. There are, however, other attitudes that can be of critical importance in determining whether or not an inspection is effective.

While it is true that some fire service organizations and individuals have limited police powers, the vast majority do not have such powers, nor should they have them. Even those organizations and individuals that do have such powers should exercise them with care. Fire safety programs succeed or fail on the basis of whether or not they receive broad public support. To achieve such support, inspectors must not lose sight of the fact that fire prevention is essentially a service operation. In other words, the effective inspector, in addition to working for his or her employer, operates on the premise that he or she is also working for the best interests of the facility being inspected.

No matter how competent an inspector is, there is no way a plant or institution can be completely inspected by one individual within an acceptable time frame. It is therefore essential for an inspector to establish a relationship of mutual trust with master mechanics, plant engineers, plant fire chiefs, and other key employees. These individuals know where the out-of-the-way hazards are located and have other information important to fire safety. Once the key employees know they can trust the inspector to handle a situation so that it will not boomerang on them, they can be of great help in making a plant safer.

Good inspectors know how to listen. The inspector may know fire hazards a lot better than plant or institutional management, but he or she does not necessarily know operating procedures better than management. Once supervisors become aware of a hazard, they can often use their knowledge of operations to help find a solution to the problem. One of the finest and most successful inspectors the author ever knew seldom made a recommendation. He had developed the art of leading discussions around to a point where management proposed essentially the corrective action he thought necessary. Contrast the following items on a report: (1) Provide _____ for the protection of _____, and (2) Plant management is planning on providing _____ for the protection of _____. Work with management—it pays.

Occasionally, a plant or institution will be found to have a number of hazards. All the hazards should be identified, but it is usually unwise to press for correction of all of them at one time. Costs being what they are, the 100 percent compliance attitude often results in management "getting its back up" and doing nothing. Rather, inspector and management should set up a system of priorities and work on the most important first. As long as the defects do not threaten life, key operations, or excessive property values, it does no harm to temporarily put minor items on the "back burner" in order to be sure of getting major items done.

In summary, an inspector should lead and not drive, suggest rather than demand, motivate rather than compel, and pull rather than push. Inspectors who follow this philosophy find that their reputation precedes them, and the people they deal with are far more ready to cooperate.

Inspection Frequencies

It is an unfortunate fact of life that, at the present time, many communities do not have enough inspectors to make thorough inspections of the various types of properties at the frequencies that are desirable. Where this condition exists, it is important to establish priorities. All inspections should be thorough, but the frequency at which they are made may be altered to fit the personnel time available. Making inspections at the desired frequency but without the personnel to do a respectable job is largely a waste of time. Management knows when it is being shortchanged.

The area that can be satisfactorily covered during an inspection depends on such features as (1) construction—number of stories and size of areas, access, and other design features; (2) occupancy—slight, ordinary, or extra hazard; and (3) protection—requirements for testing installed systems, need for flow tests and other procedures.

To set up a schedule of inspections that is workable with available personnel, one should first classify the types of buildings into a few major groups, determine the approximate square footage of coverage per hour for each type of building, and specify the desired frequency of inspections. From this information is derived the basic overall requirement for personnel, based on an assumption of average conditions throughout. Table 10 illustrates preliminary scheduling.

For each hour spent inspecting, it is reasonable to assume that half

Table 10
Preliminary Inspection Schedule

Class of building	Total Sq Ft	Inspection Rate Sq Ft Per Hr	Frequency	Total Hr Required
Administrative	1,200,000	75,000	Quarterly	64
Hospital	150,000	50,000	Quarterly	12
Industrial	22,000,000	50,000	Quarterly	1,760
Warehouses	11,000,000	100,000	Quarterly	440
				2,276
Public Assembly	400,000	100,000	Quarterly	16
Residences	10,000 Bldgs.	3 Bldgs.	Annually	3,333
				5,625

an hour is required for report writing and essential research. Therefore, the workload shown in Table 10 actually requires 8,437 personnel hours. Assuming forty-hour weeks and four weeks for vacation and sick leave, each inspector has 1,920 available working hours annually. And $8,437 \div 1,920 = 4.4$. Accordingly, five people are needed to carry the workload. If the organization doesn't have five people and wants to do an effective job, something has to give. Some acceptable ways of doing this are listed below.

Cut frequency of inspections of the smaller, low hazard risks. This is easiest to do in administrative and warehouse classifications.

Cut frequency of inspection at low to moderate value locations, unless unusual hazards are present. For example, a department might inspect values over $5 million quarterly, values of $1 million to $5 million three time a year, values of $500,000 to $1 million semiannually, and values below $500,000 annually.

After the first inspection, it may be possible at some locations to increase the permissible inspection rate. The final schedule should show frequencies that can be met without downgrading the effectiveness of individual inspections.

Preparation for Inspections

Even if an inspector has a basic background in construction, occupancy, and protection, there remain several other areas where a little advance preparation can pay large dividends.

I. No one individual can possibly remember all of the construction, occupancy, and protection fire safety requirements for the thousands of types of buildings and activities that exist. It is possible, however, to have a small but adequate reference library. These references should

Fire Prevention Inspections

be used as necessary to supplement the basic guidelines that every inspector should know. For example:

   A. Plant A produces flour. A quick review of a few pages of reference material on flour mills will provide an insight into special hazards outside the inspectors' training and experience and point the way to a solution of any problem.

   B. Warehouse B is a one-story, noncombustible, sprinklered warehouse used for the storage of slow-burning Class A materials on wood skids. Here, no research should be necessary.

   Whatever is done in the way of using reference materials, they should be left in the office or in the car. Carrying reference materials into a plant implies that the inspector doesn't know his or her job.

II. If possible, arrangements should be made to have new inspectors accompany experienced personnel for a reasonable length of time. Such training provides valuable information on how to approach different types of problems and situations and also helps in developing the judgment necessary to distinguish between major and minor problems.

III. Some organizations allow all inspection personnel to make full use of past inspection reports and allied data in preparing for an inspection; others limit the data available to field personnel. In the author's opinion, the latter are correct. Even though the overwhelming majority of inspectors make independent inspections whether or not they have past reports, some may let their judgment be overly influenced by past work, and a few are careless and depend on those before and after them to do the job right. Examples of records that should or should not be provided include, but are not limited to, the following:

   A. Water flow test records should be provided. It is important to know whether or not water supplies are holding steady. If these tests are rightly compared with one another, the presence of possible obstructions or the need to schedule improvements may become apparent. One test by itself cannot provide such information.

   B. Plot plans should normally be provided. Strictly speaking, they are not necessary, but their availability is often a time-saver. Large plants in particular are subject to frequent changes, and making sketches of additions or deletions is much easier when a master plan is available.

   C. A record of tenants (who, where, and what they do) can be very helpful in a multitenanted location.

D. Information on potentially ticklish situations should be available. There won't be many of these, but there will be some.

E. Some organizations list only major recommendations on the formal inspection report and carry minor recommendations on separate sheets. When this is done, it is usual to leave an informal list of minor recommendations with the master mechanic or another responsible person below the managerial level. If a number of minor recommendations have previously been made, the inspector should have a list of them to check the extent of compliance. Unless there is an obvious lack of cooperation, action on minor recommendations should be kept essentially informal.

F. Major recommendations on previous reports should be available to beginning inspectors. Whether or not they are available to experienced personnel is a matter of choice. Repeated tests by various organizations have proven conclusively that such records are not needed by experienced personnel.

G. Except possibly at the very beginning of his career, an inspector should not have access to ratings such as "Excellent," "Good," "Fair," "Poor," "Satisfactory," "Satisfactory except as Noted," or "Unsatisfactory" from a previous report. The fire prevention office is entitled to the independent unbiased judgment of each inspector.

In summary only that information that will be helpful to the inspector without influencing his or her judgment should be provided.

IV. Inspection preparation should include making certain that the necessary equipment is picked up in advance. It does not look good to have to borrow equipment from a plant. Consideration needs to be given to special requirements in addition to the usual gauges, flashlight, and other equipment. For example, the author carried a coil of manilla rope in the trunk of his car during the water-testing season. The purpose of the rope was to tie down hose lines for water tests at plants where help was not available to hold the hose.

V. Many ways have been suggested for taking notes during inspections. This is a matter of personal choice, but from the author's point of view a pocket notebook is the most practical. It is unobtrusive and can be used to include pertinent preinspection notes where desirable.

Inspections

Fire prevention inspections, properly handled, probably do more than any other single activity to instill consciousness of fire safety in the minds of persons with the authority to do something about it. Experience has repeatedly proven that where there is a good inspection program fire losses drop sharply. Every effort should therefore be made to have a successful fire prevention inspection program.

Even if inspectors have an adequate technical background, an effective understanding with management and appropriate inspection preparation, there are still a number of ways in which inspections can be made even more effective.

I. Rotation of inspectors. Larger communities and larger private and public fire prevention organizations generally have a number of inspectors. Many of these organizations make a point of hiring personnel having various backgrounds. All have the same basic background, but one may also have special knowledge of construction, another of electrical equipment, another of ovens, and so on. If inspection of a plant is rotated among several such inspectors, a good general fire prevention inspection results every time, and, in addition, each inspector brings to bear his own specialized knowledge during his tour of inspection.

Smaller organizations do not have the internal capacity to provide specialized services in all areas, but they can develop alternate means of doing the same thing. First of all, they have or should have basic fire prevention capabilities that cover most of the situations that may arise. For special problems, there are two basic sources of additional necessary information.

A. A small reference library, tailored to the needs of the community served.

B. Contact with organizations and/or personnel with the expertise needed. Occasionally, an inspector working alone is hesitant about asking for advice from a national organization or a well-known specialist. There is no need for such hesitance. These organizations and personnel are as interested as the lone inspector in cutting fire losses and, within reasonable limits, will extend a helping hand.

II. Scheduling times for visits.

A. Unannounced inspections. If where it is possible—and in most instances it is—arrangements should be made in advance for unannounced inspections. The author has found that management, in general, appreciates the fact that unannounced inspections are usually more effective, for obvious reasons. The privilege of unannounced inspections, however, must be exercised with discretion. For example:

1. Security. Some plants, laboratories, and offices handle classified materials, including national security items. Where such conditions exist, advance notice must be given so that sensitive mate-

rials can be covered or otherwise secured. If the sensitive items form real fire hazards, it becomes the inspector's responsibility to obtain the necessary security clearance.

2. Retooling. Many plants shut down for two or three weeks in the summer for retooling and maintenance that cannot be done during operating periods. At such times, the plant is short-handed, and people that the inspector should talk with are already saddled with a heavy workload. Visits at such times are seldom appreciated.

3. Seasonal workloads. Certain occupancies, such as department stores and some types of clothing manufacturing, have short periods of extremely heavy workloads when everybody is working at capacity. Visits to such locations should be scheduled a reasonable time before peak periods, for several logical reasons. For example pre–peak load inspections give a plant time to correct deficiencies before the peak period begins.

4. Emergencies. On some occasions, a plant may be faced with an emergency situation that has nothing to do with fire. Inspections during such periods do not make for good relations.

B. Off-hour inspections. Although most inspections are made during normal business hours, there are situations when an inspection should be made during off hours. Permission for such inspections should always be obtained in advance. Two situations that may indicate off hour visits are described below.

1. It is unfortunately true that, all too frequently, night watchmen and guards are elderly, physically incapable of handling an emergency, and poorly trained. If such a situation is suspected, it is wise to diplomatically arrange to make a night tour with the watchman and a management representative or to have the watchman held over for an interview in the morning. This type of situation requires the exercise of tact.

2. Night shift operations may be safer or more hazardous than daytime operations because of differences in the degree of supervision or in operating requirements. The only way to find out is to make an occasional night check, first obtaining management permission.

C. Identification. At each visit, inspectors should identify themselves properly and ask permission to inspect. During the inspection, a representative of management (usually the plant engineer or master mechanic) should accompany the inspector. In some loca-

Fire Prevention Inspections 299

tions, particularly in union shops, a labor representative may also go along. In no case should an inspection be made unaccompanied.

III. Moral responsibilities. Some inspectors are responsible for checking both life and property hazards; others are responsible for checking property hazards only. From an official standpoint, the latter group can only comment on fire hazards to property. But for the most part, they are also quite capable of picking up serious life hazards. Officially their hands are tied, but, in the author's opinion, they have a moral responsibility to point out such situations and unofficially suggest the plant give the matter some study. An inspector can easily do this without assigning any responsibility to his employer.

IV. The tour.

A. Since there is no such thing as a standardized plant, the method of making a tour must necessarily be adjusted to fit the particular situation. However, to inspect a fairly large facility with numerous buildings in the same general area, it is usually desirable to follow approximately the procedure outlined below.

1. Exposure by other property. Quite often it is possible to inspect this prior to starting the actual property inspection.

2. In many cases, it is desirable to begin an inspection with a quick tour of the yard, or for a very large plant, a section of the yard in order to fix the relationships of the buildings to one another, to yard operations, and to access routes for fire departments.

3. Where buildings are of fairly uniform height, it is desirable to proceed next to the area where raw stocks are received and then follow through to finished products. In most plants, however, buildings differ greatly in height. In such situations, it is usually best to start on the roof of the tallest building to get a clear view of the entire plant and then work down through this building. Additional buildings should be inspected in systematic order, usually starting from the top and working down. Whatever method is used, it is important to be systematic and leave no area unvisited.

4. When entering an area, it is wise to stop for a moment and make a quick visual survey. It is possible to train one's eyes to pick up quickly a great deal of information regarding construction; extent of installed protection, including valve locations; general housekeeping standards; locations of hazardous operations; and the like. From this survey, an inspector can plan his route so that

no important area is missed. It takes time to develop this ability, but it should be done. It is important to see an operation, not only as a series of parts, but also as a whole.

5. In traveling through a plant, an inspector must be alert to potential construction and occupancy hazards such as those outlined in previous chapters. These hazards or deficiencies should be discussed as appropriate with personnel accompanying the inspectors so as to obtain their ideas and some preliminary information about whether or not any corrective action is programmed.

6. Some deficiencies are inevitably found in any operating plant. In dealing with them, an inspector—to borrow a timeworn phrase —needs to learn to separate the sheep from the goats. If this is not done, the final interview and the final report will lose a great deal of their effectiveness.

   a. At any operating plant, it is quite possible to find an occasional cracked wall outlet, a slightly leaking safety can, a missing guard on a portable light, an extinguisher overdue for a recharge, and the like. Unless there are many deficiencies of a similar nature, it is usually possible to have the deficiency corrected on the spot or to arrange for its prompt correction. Top management's time should not be taken up with a few minor defects.

   b. If there are numerous minor deficiencies, the inspector should record them in duplicate, giving one copy to the escort and retaining a copy for the next inspector's use. In the report to top management, these minor recommendations should be consolidated into a single statement, such as "The lubrication schedule for bearings of loom motors in Weave Sheds 2 and 3 should be revised to provide proper lubrication (30 percent of the loom motors checked had underlubricated bearings)."

7. During each inspection, the inspector should make certain that all installed protection systems are operable and have such tests made as are necessary to prove that the systems function properly. Under no circumstances should an inspector operate any valve or test any equipment. He or she is there to observe and give advice and for nothing else.

   a. Precautions.

      1. Closed valves. Find out why they are closed before asking for them to be opened. It could be embarrassing to find out that the system is being worked on and there are open-end

pipes. If there is no legitimate reason for the valve to be closed, have it opened and find out why it was closed.

2. Supervision. The majority of sprinkler protected plants and many with fire detection systems have central station or fire department supervision. No tests should be made without first notifying the supervising agency. After testing is completed the supervising agency should be notified and requested to confirm alarms received. Any failures should be checked out immediately.

b. Some common kinds of tests.

1. Tanks. An inspector should check the water level of both gravity and pressure tanks and the air pressure of pressure tanks. In cold climates gravity tank heating systems should be checked and the tanks inspected for ice. In most cases, it is not necessary to climb long ladders to ascertain the necessary facts.

2. Drain tests.

3. Dry pipe valves. Check for air pressure, heat in valve house in winter, water columning, and other features. Also check on annual trip test.

4. Flow tests. Plants with hydrants and yard systems or with pumps equipped with hose headers should be requested to conduct flow tests annually. When flow tests are conducted, each supply should be tested separately and then all supplies together. When tests are completed, make certain all control valves are open and sealed.

5. Check other systems as deemed necessary.

6. Check plant self-inspection forms. It is reasonable to be skeptical of clean inspection forms, all neatly filled out and kept in an office desk drawer. It is difficult to see how an inspection report form supposedly carried through a plant can wind up perfectly clean and with no finger marks. As a case in point, the writer remembers a plant that had "daily" control valve checks—according to papers filed with the superintendent. Yet, at 10:30 one morning, this writer found a 6-in. OSY valve closed and covered with dust within 25 ft. of the superintendent's office. The plant inspector's report for that same day stated that at 10:00 A.M. the valve was open. How long that important valve had been closed is a good question. For their own protection, plant inspectors should take the

forms with them, fill them out as they check, and keep these originals for visiting inspectors.

7. If there is a watchman, check clock dials for completeness of rounds at proper intervals. Learn how to detect from the dials how many times the clock has been opened. The inspector should also know if the watchman changes his own dials (this is undesirable practice). If there are key stations, a spot check should be made on how well the key chains are fastened to the wall. It is not unknown for an enterprising watchman to file links so that he can remove the keys on the first round, punch the middle rounds while sitting at a comfortable desk, and put back the keys on the last round.

8. Inspection tour procedures as outlined in proceding paragraphs should be modified as necessary to fit institutions, hotels, small plants, and other operations. Fundamental principles should, however, remain the same.

9. Last but not least, life safety should be taken into account. Inspectors with life safety responsibility should be prepared to work with management to correct deficiencies. Other inspectors would be remiss in not pointing out serious life hazards, but they normally would be out of line in recommending a specific remedy.

Exit Interviews

Except in small facilities, it is rare for management officials with major decision-making authority to accompany the inspector. Top management has many major responsibilities, and the time available for any one subject is limited. For this reason, two exit interviews are recommended.

I. The first exit interview should be with the plant engineer, the master mechanic, or some other management representative. At this time, it is appropriate to discuss in detail findings other than those that have already been corrected. Recommendations that are within the management representative's authority to carry out should be cleared first. If the interview is handled properly, it is usually possible to reach a meeting of the minds on appropriate corrective actions. All the inspector then needs to do about these recommendations is to make a brief note for the next inspector's use. The final step in the preliminary exit interview is to discuss those recommendations that are beyond the escort's authority to carry out. Again, it is usually possible for inspector and escort to reach an agreement in principle on proposed corrective actions. The management representative has a right to know in advance what recommendations affecting his or her operations the inspector is planning to discuss with top management.

II. The final exit interview should be kept brief and to the point.

A. Recommendations, if any, and the reasons therefore should be thought out in advance and phrased so as to be readily understandable to non–fire safety personnel. The inspector should also be prepared to listen to any alternative methods of handling the situation. What the inspector is after is fire safety, not a particular method of doing something.

B. The facility may be in the process of adding or removing buildings, installing a protection system, changing occupancies, or other alterations. Management comment on such changes is desirable and usually easily obtainable, provided the inspector restricts his or her questions to really important points.

C. If minor items have been cleared earlier with subordinates, it is in order to mention that several minor items were discussed with the plant engineer, who is taking care of them. Normally, top management does not want details on such matters.

D. Always close an interview with an expression of thanks for the courtesies provided during the inspection.

Reports

Every inspection should be followed up by a written report to the property owner. The amount of detail required varies with the type of report, as indicated below. However, all reports, regardless of their type, should have one feature in common: nothing should appear in the narrative parts of the report that was not discussed with management.

I. Candidate, or initial, reports are those made as a result of the first inspection of a property. They generally contain considerably more information than will following reports. For example, an insurance company needs to know a great deal about the extent of potentially hazardous occupancies, even though they are properly safeguarded, in order to establish insurance rates. A detailed outline of construction, exposure, and protection features is needed for the same purpose. Most property owners are aware of the need for a detailed report on the first visit and will cooperate. It is wise, insofar as occupancy descriptions for manufacturing plants are concerned, to outline to management what is proposed for inclusion in the report. It is possible at times to inadvertently let out a trade secret unless this is done. In the narrative part of the report should appear recommendations, if any, and any necessary comments on proposed future changes of importance. While report formats differ, candidate inspection reports, in addition to the narrative, should include the following information.

    A. Date(s) of inspection.

    B. Name of inspector.

    C. Name and address of property.

    D. Names and titles of persons interviewed.

    E. Names of principal tenants (can be in narrative).

    F. Occupancy class (for example, machine shop, textile, office).

    G. Plot plan of building(s) (showing construction, protection, and other relevant features).

    H. Protection details (for example, percent sprinklered, percent needing sprinklers).

    I. Construction details (for example, percent fire resistant, percent noncombustible, percent brick joisted).

    J. There should also be provision for rating such items as housekeeping, maintenance, watchman service, exposure, water supplies, and valve and alarm supervision. (Common practice is to use ratings of (1) satisfactory, satisfactory except noted, and unsatisfactory or (2) excellent, good, fair, and poor.)

II. Regular reports are for those complete inspections following an initial report. Rating blocks for these reports are the same as those for candidate reports. There is no accompanying plot plan unless there were building changes between inspections. In the narrative portion, general property descriptions are dropped, and the narrative is limited to the following.

    A. Recommendations, if any.

    B. Important recent changes, if any.

    C. General remarks, to convey such mundane information as the location of particular tenants. There is seldom need for a general remarks section in a report on owner-occupied property.

It is important not to make regular reports overlong. It is safe to say that, if inspections and interviews are handled properly, the reports on 90 percent of regular inspections can be put on a single report form sheet. In fact, many reports can be completed by filling in the rating block and using only the face side of the form for comments.

III. Special reports are those made to cover a particular problem. They may cover a water supply investigation, a review of flammable liquids operations, or an inspection of any other operation involving fire safety. Because of the wide variety of special reports, there is no general format. Such reports should be limited to covering thoroughly the operation in question.

IV. Life safety reports may be included as part of candidate or regular reports or be completely separate. Either way, the reports should include information on adequacy of exits, requirements for fire drills, areas of refuge where needed, requirements for attendance around the clock (for example, in nursing homes) and any other provisions essential to life safety. Formal reports covering life safety only should omit details that are property oriented and do not affect life safety. It should be noted, however, that the life safety inspector has a moral obligation to unofficially point out any major property hazard.

Summation

In the final analysis, it is the caliber of the inspections and the reports of these inspections that determine how effective a fire prevention program really is. Nowhere else do fire prevention personnel have the opportunity to come in such close personal contact with both management and employees. Nowhere else is there a better chance to make the places where people live, work, and play safer from fire. The opportunity is there and all fire prevention personnel should train themselves to use it to best advantage.

Appendix

Appendix A—Metric Conversion Table

1 inch = 2.54 centimeters
1 foot = 0.3048 meters
1 yard = 0.9144 meters
1 mile = 1,6093 kilometers
1 square inch = 6.4516 square centimeters
1 square foot = 0.0929 square meters
1 square yard = 0.8361 square meters
1 square mile = 2.590 square kilometers
1 cubic inch = 16.387 cubic centimeters
1 cubic foot = 0.0283 cubic meters = 28.32 liters
1 cubic yard = 0.7646 cubic meters
1 kilogram = 2.205 pounds
1 metric ton = 1,000 kilograms = 2,205 pounds
1 kilogram per square centimeter = 14.22 pounds per square inch
1 U.S. gallon = 3,785 liters
1 liter per second = 15.85 gallons per minute
Temperature centigrade = 5/9 (temperature Fahrenheit −32°)

Appendix B—Other Useful Measurements

1 acre = 43,560 square feet
1 square mile = 640 acres
1 cubic foot = 7.48 U.S. gallons
1 cubic foot per second = 448.3 gallons per minute
1 kilowatt = 1,000 watts = 1.34 horsepower
1 U.S. gallon of water = 8.33 pounds
1 foot of water = 0.433 pounds per square inch = 0.881 inches mercury

Appendix C—Useful Formulas

C = circumference A = area r = radius
d = diameter π = 3.1416 V = volume
Circumference of circle $C = 2\pi r = \pi d$
Area of circle $A = \pi r^2 = \pi d^2/4$
Volume of sphere $V = 4\pi r^2/3$
G = gallons per minute p = pressure in pounds per square inch
h = head in feet of water v = velocity in feet per second
d = diameter c = coefficient of discharge
Discharge $G = 29.83\ cd^2 \sqrt{p}$
Velocity head $h = V^2/65.4 = G^2/384d^4$

Selected Reference Bibliography

Construction
    Brannigan, Francis L. *Building Construction for the Fire Service.* Boston: National Fire Protection Association, 1971.

General
    Factory Mutual System. *Handbook of Industrial Loss Prevention.* 2d ed. New York: McGraw-Hill, 1967.
    National Fire Protection Association. *Fire Protection Handbook.* 14th ed. Boston: National Fire Protection Association, n.d.
    _____. *Inspection Manual.* Boston: National Fire Protection Association, 1970.
    _____. *National Electric Code.* Boston: National Fire Protection Association, n.d.

Hazardous Materials
    Bahme, Charles W. *Fire Officer's Guide to Dangerous Chemicals.* Boston: National Fire Protection Association, 1972.
    Meidl, James H. *Flammable Hazardous Materials.* Beverly Hills, Calif.: Glencoe Press, 1970.
    National Fire Protection Association. *Fire Protection Guide on Hazardous Materials.* Boston: National Fire Protection Association, 1975.

Investigations and Inspections
    Kirk, Paul L. *Fire Investigation.* New York: John Wiley & Sons, 1969.
    Robertson, James C. *Introduction to Fire Prevention.* Beverly Hills, Calif.: Glencoe Press, 1975.

Index

Acetyl chloride, 187
Acetylene, 187; commercial uses of, 217, 219; generation, 177–78
Acid anhydrides, 187
Acids: commercial uses of, 217; corrosive, 196–97
Acrylonitrile liquid, 202
AGA Directory of Certified Gas Appliances and Listed Accessories, 18
Agencies, supporting, 283
Air- and water-reactive chemicals, 186–89
Air conditioning, 34, 66, 256; shutoffs, 135
Air flow, controlling, 76–77
Air-handling systems, 75–82, 133–40
Air intakes or outlets, 77, 134–35; flammable vapors and, 139–40
Air pressure tests on sprinkler systems, 105–6
Air-supported structures, 46
Alarms, delayed, 275–76
Aldehydes, commercial uses of, 217
Alkali metals, 187–88
Alkalis, 197–98

Alkaloids, 200
Alkyl boranes, 187
Alterations, hazards of building, 29–30
Aluminum, oxidation of, 190
Aluminum alkyls, 187
Aluminum phosphide solid, 202
America, fire safety in colonial, 3
American Gas Association, 18
American Insurance Association, 18
Ammonia compounds, commercial uses of, 217
Ammonium nitrate, 192
Aniline, 200
Antimony compounds, 200
Arkansas, train wreck in (1960), 192
Armed forces, U.S., 13
Armored cable, 132
Aromatic hydrocarbons, 200
Arsenates, 200
Arsenites, 200
Arson: losses from, 221–24; motivations for, 224–27; versus nonarson investigations, 264
Arson law, 227–31
Askarel-insulated transformers, 68
Association of American Railroads, 18

309

310 INDEX

Atropine, 200
Autoignition temperature of flammable liquids, 155–56
Automatic sprinklers, 98–109; in ovens and driers, 148
Autopsies in fire deaths, 230–31
Auxiliary fire extinguishing systems, 124
Auxiliary utility systems, 65–66
AZO compounds, commercial uses of, 217

Ballasts, light, 131
Balloon frame buildings, 26
Benzene, 187; liquid, 202
Blind spaces, 34–35
Boiler rooms, safety of, 91, 92
Boiling point of flammable liquids, 158
Boston, night club fire (1942), 5, 22; fire prevention ordinance of 1631, 3
Breaker tieback, 72
Brick arches in construction, 25–26
Bromates, 191
Bromine, 194–95
Building codes, 22–24; exceptions to, 24–25; interior finish requirements of, 36; wiring provisions of, 73
Building construction, 21–47; changes in types of, 3–5, 44
Building deficiencies, 8–9
Building occupancy, 4–5, 41–42, 109, 125–26, 251–62
Buildings, classes of, 24
Bulk cylinder storage, 175–76
Bureau of Explosives, 18–19
Bureau of Mines, 14
Burning intensity rate, 54
Butyl compounds, commercial uses of, 217

Cable troughs, 130
Calcium, 187–88
Calcium carbide, 187
Calcium chloride, 187
Calcium oxide (quicklime), 197
California, office of the fire marshal of, 205
Canopies for cylinders, 175
Carbides, 187
Carbon dioxide, cyanides and, 201
Carbon dioxide protection systems, 116–20
Carbon disulfide liquid, 202
Carbonic acid, 201
Carborundum, 187
Carpeting, 37

Cast iron building frameworks, 25
Casualties, reasons for fire, 276–77
Caustics, 197–98
Ceilings, dropped, 33–36
Chemicals: corrosive, 195–99; toxic, 199–205; unstable, 216–20
Chicago theater fire (1903), 5, 22
Chimneys, 94–96
Chlorates, 191
Chlorides, 187
Chlorine, 194–95; compounds of, 217
Circuit breakers, 71, 74, 132
Civic organizations, fire prevention education by, 288
Clearances, chimney, 95
Cleveland, LNG explosion in (1944), 174
Coal safety, 90
Coast Guard, 13
Cocaine, 200
Cocoanut Grove nightclub fire (Boston), 5, 22
Combustibles, control of, 93
Combustibility of the environment, 8
Compensator tiebacks, 72
Computer operation in industry, 255
Concealed spaces, 33–36
Concrete, introduction of, 4
Conduction, 50
Conflagrations, 56; defenses against, 57; outside, 260
Construction: hazards of, 28–30; major types of, 24–28; modern, 26–28
Control devices, electrical, 72–73
Controls, air flow, 75–77
Convection, 50
Conversion programs, gas, 89
Conveyor openings, 32
Cooling towers, 79–81
Corrosive chemicals, 195–99
Crawl spaces, 36
Creosols, 200
Critical components in industry, 256
Cutting and welding, 29; hazards of, 178–79; as ignition sources, 273–74
Cyanides, 201
Cyanogens, 201
Cylinder containers: for liquified natural gas, 174–76; for toxic chemicals, 201–2
Cyrogenic liquids, 212–16

Damage, estimating fire, 277–79
Day-care centers, 43
Deaths: autopsies in fire, 230–31;

Index

causes of fire, 276–77
Decatur, Illinois, fire in (1974), 172
Decomposition, violent, 218
Defenses against exposure, 57–60
Deflagration, 218
Deluge sprinkler systems, 104
Department of Defense, 13
Department of Transportation, 13–14, 174
Detonations, 218
Diborane, 187
Dip tanks, 164–65
Disasters: involving liquified petroleum gas, 172; major fire, 5–6, 22; recovery from, 258–60
Double powder foam generation, 114
Driers, industrial, 145–50
Drum storage, indoor and outdoor, 161–62
Dry chemical extinguishing systems, 120–22
Drying racks, combustible, 93
Dry pipe sprinkler systems, 101–3
Dry transformers, 68
Duct systems, 78–79; multibranch, 139
Ductwork, building system, 136
Dust collectors, 137–38
Dust explosions, 129–30, 247; prevention of 136–38

Eagle Pass, Texas, fire in (1975), 172
Earthquakes, fires caused by, 259
Economic facts of fire prevention, 12
Education for fire prevention, 284–89
Electrical equipment: defects of, 130–31; investigation of, 267; in oven driers, 147; portable, 131–32
Electrical ignition, 270–71
Electrical systems, 66–75, 126–35
Electroplating, 199
Elevators, 65
Empire State Building, 44
Escalators, 31–32
Escape route, presurveyed, 24
Espionage incendiary fires, 225
Ethers, commercial uses of, 218–19
Ethylene oxide gas, 202
Etiological agents, 203–4
Evacuation areas in skyscrapers, 45
Evidence in arson cases, 235–38
Exhaust safety ventilation, 136
Exit interviews on inspection tours, 302–3
Exitways, 30–31
Expansion joints, 32

Explosions: industrial, 205–12; venting, 160
Explosive range of flammable liquids, 154–55
Explosives, commercial, 205–12
Exposure, investigation of, 268
Exposure fires, types of, 55–57
Extension cords, 74, 132

Factory Insurance Association (FIA), 17
Factory Mutual System (FM), 17, 85, 169, 251; *Approval Guide*, 17
Fail-safe systems for furnaces, 142
Fall River, Massachusetts, drier fire in (1941), 149
Fan motors, protective devices for, 77
Federal Aviation Administration, 13
Federal government, fire preveniton organizations of the, 12–14
Federal Highway Administration, 13
Federal Railroad Administration, 13
Fiber board, low-density, 38; compared to particle and hardboard, 38
Filters, on air handling systems, 81, 135
Finishes, interior, 36–38
Fire: barriers, 62; causes of in ovens and driers, 147–48; delayed discovery of, 275; external exposure, 49–60; point of origin of, 269
Fire and Air War, 57
Fire Command, 16
Fire control: cost effectiveness of, 6; 19th century, 4–5
Fire cutoffs: adequacy of, 92–93; horizontal and vertical, 30–33, 59
Fire Journal, 16
Fire losses: indirect, 278–79; statistics on, vii–viii
Fire marshals' offices, 15
Fire patterns, in arson, 231–33
Fire Prevention Code, 18
Fire Prevention Guide on Hazardous Materials, 202
Fire Prevention Handbook, 17, 128
Fire prevention: legislation, 2–3; seminars on, 288; services, definition of, 9
Fire Prevention Week, 281, 284
Fire protection: during construction, 28–29; private, organization, 9, 16–19
Fire pumps, electrically driven, 132–33
Fire storm, 56–57
Fire Technology, 16

Fire tests: on construction materials, 39–41; on plastic, 248
Firing procedures, safety of, 91–92
Flame spread ratings, 40–41
Flammable liquids and gases, 29, 138–40, 153–67, 268, 271–72; flash point of, 154
Floods, fires caused by, 258–59
Floor leakage, 50
Floors, wiring under raised, 133
Fluorides, 201
Fluorine, 194–95; 219; liquified, 215–16
Flushing methods for sprinkler systems, 108–9
Foamed or expanded plastics, 246–47
Foam protection systems, 112–16; as backup for dry chemicals, 121
Forest service: state, 15; U.S., 14, 15
Forrestal, explosion on the (1967), 208
Friction ignition sources, 273
Fuel safety, 90–91
Fuel systems for furnaces, 142–44
Fulminates, commercial uses of, 218
Fumigants, 202–3
Furnaces, industrial, 141–45
Fuse, 72; boxes, 74

Gas laws, basic, 83–84, 168–69, 212–13
Gas leak ignition, 273
Gas mains and piping, 34–35, 66, 87–89
Gas systems, 82–89; for furnaces, 143
Gases: flammable, 138–40, 167–79; noncompatible, 175; storage facilities for, 84–86, 91; unburned, 95–96
Gasoline-powered equipment, 29
General Motors fire (Livonia, Michigan, 1953), 6–7, 256
Generators: inside, 111; types of, 68–69
Germ agents, 203–4
Germany, nitrate explosion in (1923), 192
Glass windows, wired, 58–59
Grouped cables, 71–72

Halogens, 122–23; use and storage of, 194–95
Halon 1211 and 1301, 122–23
Hammurabi, code of, 1
Handbook of Industrial Loss Prevention, 17, 128
Hangers, sprinkler, 108
Hazard classes, 128
Hazards: of electrical equipment location, 126–28; sprinkler systems and, 101; wiring, 74
Hazardous Materials Regulations Board, 13–14
Hazardous operations, control of, 29–30
Heat, oxides and, 191
Heating plants, 256
Heating systems, 89–96, 272–73; control deficiencies of, 91–92
Heat treating, 193–94
Hexane, 187
High-rise buildings, safeguards for, 44–46
Highways, buildings over, 46
Home, fire prevention in the, 286–87
Hospitals, life safety requirements for, 42–44
Housekeeping as a fire prevention measure, 251–54
Hurricanes, fires caused by, 259–60
Hydrides, 187
Hydrochloric acid, 196–97
Hydrofluoric acid, 196–97
Hydrogen, 187; liquified, 214, 216
Hydrogen chloride, 187; vinyl produced, 247
Hydrogen cyanide, 199, 202; gas, 201
Hydrogen peroxide, 219
Hydrogen sulfide, 188
Hydrostatic tests for sprinkler systems, 105
Hypochlorite, 191

Ignition sources: eliminating, 160–61; proliferating, 8; types of, 269–74
Incendiarism, 225
Incinerators, industrial, 150–51
Industrial chemicals, 185–220
Industrial fire prevention, 19, *passim*
Industrial gases, properties of, 82–84, 169–70
Industry, fire prevention education by, 289
Inerting, 161
Inflation factor in fire losses, 222
Infrared lamps, 166–67
Inspection and rating bureaus, state, 15
Inspection programs, 291–305
Institutions, fire prevention education in, 287–88
Insurance companies: fire insurance policies of, 22; formation of, 3; loss payments by, 277–78
Insurance Grading Schedule, 18
Insurance Services Offices, 18

Index

Internal exposures, 60–63
Interior finishes, 36–38
Investigative techniques, 267–79
Investigation, arson, 229–35; fire loss, 263–79
Iodine, 194–95
Iroquois Theater fire, Chicago (1903), 5, 22

Jet propulsion fuels, 157
JP4 fuel, 157

Lectures and demonstration on fire prevention, 284
Life Safety Code, 16
Lighting-off procedures, improper, 91
Lights, portable, 74; suspended, 131
Lint fires, 130
Lithium, 187–88
Liquid propellants, 210–11
Liquids: classifying, 158; flammable, 138–40, 154–67, 268
Liquified natural gas (LNG), 82, 172–74; storage, 85–87
Liquified petroleum gas (LPG), 82, 170–72; storage, 85
Loan shark fires, 223–24
London, England, building ordinances of, 2–3
Loss Prevention Data Books, 17
Low pressure carbon dioxide system, 118

McCormick Place fire, Chicago, 22
Maintenance: of carbon dioxide systems, 120; of dry chemical systems, 121–23; of foam systems, 116; of Halon systems, 123; procedures for heating systems, 93; of sprinkler systems, 106–7
Manifolds for cylinders, 175
Mare's nests, 131
Masonry walls, 58; lack of tension strength in, 40
Mass media, public relations use of, 282–83
Massachusetts, gas explosion in, 177
Mechanical foam extinguishing systems, 115
Metal grinding, 138
Methane, 187
Methyl bromide gas, 202
Middle Ages, fire prevention in the, 2–3
Miranda Warning, the, 229–30
Mixing operations for liquids, 163–64
Morphine, 200

Mortar, chimney and stack, 95
Motors: dirty, 131; electric, 68–70
Municipal fire prevention activities, 15–16
Multibranch duct systems, 139
Multipass driers, 111
Multiple outlets, 74

National Building Code, 18
National Bureau of Standards, 14
National Electric Code, 16, 73, 128
National fire codes, 16
National Fire Prevention and Control Administration, 12–13
National Fire Protection Association (NFPA), 14, 16–17, 85, 169, 251; *Standard 13* sprinkler rule book, 109; *Standard 901* report forms, 265–66
National Park Service, 14
New England, hurricane destruction in, 259–60
New York airport terminal fire, 248
New York City, Triangle Shirtwaist Factory fire (1911), 5
Nitrates, 191; use of in salt baths, 193–94
Nitric acid, 196–97
Nitrocellulose, 240–42
Nitro-compounds, commercial uses of, 218
Noncompatible particles, 138
Non-English speaking groups, fire prevention education among, 287
Nursing homes, life safety requirements for, 42–44

Occupancy: and building sprinkler system changes, 109; effects of changes in, 41–42; fire loss and, 4–5; problems caused by, 251–62
Occupational Safety and Health Administration, 14
Oil-insulated transformers, 67–68
Oil storage and piping, 90
Oil systems for furnaces, 143–44
Outdoor pads for cylinders, 175
Outside air intakes, 134–35
Ovens, industrial, 145–50
Oxidizing agents, 190–95
Oxygen: liquified, 215; use and storage, 193

Paint, 37; combustible, 34; driers, 166–67; spraying, 165–66

Parathion pesticide, 204
Parliament, first fire prevention legislation passed by, 3
Particles, combustible, 138
Per-acids, commercial uses of, 218
Permanganates, 191
Peroxides, commercial uses of, 218, 219
Persulfates, 191
Pesticides, 204–5
Phosgene, 187
Phosphorous, 188
Physical transmission, 50
Piercing, wall, 40
Pilots for furnaces, 142
Pipe chases, 32
Pipe valves, dry, 108
Piping systems: for flammable gases, 176–77; for gas, 34–35, 87–89; for liquids, 163; steam and hot water, 34
Plant inspection, 263–64
Plant management, 260–62
Plant shutdowns, 257–58
Plastics: characteristics and uses of, 239–40; finished products, 246–48; hazards of creating, 243–46; melting of, 247
Plastic sheet finishing, 38, 288
Plywood finishes, 37
Polymerization, violent, 218
Potassium, 187–88
Potassium, carbide, 187
Potassium hydride, 187
Potassium hydroxide (caustic potash), 197
Powerhouses, 70–71
Preaction sprinkler systems, 103–4
Prealarm facilities for carbon dioxide systems, 118
Pressurized escape routes, 24
Process cooling, 134–36
Process furnaces, 142–45
Propellants, rocket, 209–12
Protection systems, 97–124; for ovens and driers, 148
Public apathy on fire prevention, 8
Public relations for fire prevention education, 282–84
Pumping systems for liquids, 163
Pyromaniacs, 225
Pyroxylin, commercial uses of, 219

Quinine, 200

Radiation exposure, 50

Rail lines, construction over, 46
Rating block, the, 266
Reader's Digest, 251, 283
Refrigerators, 133
Regulators, gas, 86–87
Report forms on fire loss, 265–67
Reports, inspection, 303–5
Research deficiencies, 9
Restraints, built-in, 38
Return air intakes, 135
Richmond, Indiana, fire (1968), 209
Rocket propellants, 209–12
Rolling steel fire doors, 30–31
Rome, fire brigades in early, 1–2
Roofing mops, 29
Rooms, concealed, 35–36
Roseburg, Oregon, explosion (1959), 208
Roseville, California, explosion, 208

Safeguards: against internal exposures, 62–63; for flammable liquids, 158–61
Salt baths, 193–94
San Francisco earthquake (1906), 259
Santa Ana winds, fires caused by, 276
Schools: fire prevention programs in, 286; fires in, 224; life safety requirements for, 42–44
Self-contained breathing apparatus for carbon dioxide systems, 118
Self-contained stored solution foam generation, 114
Shipping containers for toxic gases, 201–2
Silicon carbide, 187
Single powder foam generators, 114
Site preparation for construction, 28, 274–75
Siting of ovens and driers, 145
Skyscrapers, 44–45
Smoke damage, 50
Smoke detectors, 77
Smoke stacks, 94–96
Smoke towers in high rise construction, 45
Smoking, cigaret, 253–54
Sodium, 187–88
Sodium carbide, 187
Sodium cyanide, 199
Sodium hydride, 187
Sodium hydroxide (caustic soda), 197
Soffits, 33–36
Solid propellants, 210
Spaces, concealed interior, 33–36

Index

Spare capacity in heating systems, 93–94
Spark arrestors, 95
Special atmosphere furnaces, 144
Specific gravity in liquids, 156–57
Splices in wiring, 74
Spontaneous heating of flammable liquids, 156
Sprinkler heads, 107
Sprinkler systems, 58–59, 78; automatic, 98–109; and auxiliary systems, 124; as backup for dry chemicals, 121; cool weather valves on, 101; early, 4; in earthquake areas, 259; false alarms and, 100–101; invention of, 97
S. S. *Grandcamp*, fire on (1947), 5
Standard Schedule for Grading Cities and Towns, 18
State governments, fire control agencies of, 14–15
Static electricity, plastics and, 247
Storage: of explosives, 208–9; of flammable liquids, 159, 161–63; of gases, 84–86; of liquified natural gas, 174–76
Structural life span, 23
Strychnine, 200
Sulfuric acid, 196–97
Switchgear, 70, 71–72

Tamdem cords, 74
Tankers, 173
Tar kettles and roofing mops, 29
Technological revolution, 5–6
Telephone switchboards, 132
Temporary building, defects of, 28
Testing and maintenance procedures: for heating systems, 93; for sprinkler systems, 105–6
Tests, fire inspection, 301–2
Texas City, Texas, ship fire (1947), 5; nitrate explosion, 192
Textiles, plastic, 247–48
Thermal cutouts, 135
Thermoplastic materials, 242–45
Thermosetting plastics, 242–43
Tile arch construction, 25–26
Toluene, 187
Toxic gases, 50, 247
Transfer methods for liquids, 163
Transformers, 67–68, 132

Transportation: of explosives, 207–8; of liquified natural gas, 177
Trash disposal, 29
Triangle Shirtwaist Factory fire, New York City (1911), 5
Two-inch drains, tests of, 105

Underwriters' Laboratories, Inc., 17–18
Utility systems, building, 65–96; occupancy of, 125–40

Valves, sprinkler, 107–8
Vandalism and arson, 225
Vapor: density, 139, 157; pressure, 157–58
Vaporizers, gas, 86–87
Ventilation: for flammable liquids, 160; of ovens and driers, 146–47; positive or natural, 140; for vapors, 157
Venting devices, 62–63
Vertical and horizontal sprinkler systems, 102–3
Vinyl compounds, commercial uses of, 218

Walls, furred out, 33–36
Waste materials, 252
Water-reactive chemicals, 186–89
Water-sensitive products, 256
Water solubility of fuels, 158
Water spray system, 111–12
Water supply, 255, 276
Water tanks in earthquake areas, 259
Welding, hazards and precautions for, 178–79
Wet pipe sprinkler systems, 100–101
Windstorms, fires caused by, 259–60
Windows in high rise construction, 44–45
Winecoff Hotel fire, Atlanta, Georgia, (1946), 37
Wiring, 34; circuits, 73–75; temporary, 29; under raised floors, 133
Wood blocking, 131

Yard systems for sprinklers, 108–9
Youth groups, fire prevention education in, 287

Zoning of air-handling systems, 81–82